CHEMICAL ANATOMY OF THE ZEBRAFISH RETINA

by

Stephen Yazulla
Stony Brook University

Keith M. Studholme
Stony Brook University

Robert E. Marc
University of Utah School of Medicine

David Cameron
SUNY Upstate Medical University

A Special Issue of
JOURNAL OF NEUROCYTOLOGY
A Journal of Cellular Neurobiology
Volume 30:7 (2001)

KLUWER ACADEMIC PUBLISHERS
Boston / Dordrecht / London

JOURNAL OF NEUROCYTOLOGY

A Journal of Cellular Neurobiology

Volume 30, No. 7, June 2001

Special Issue: CHEMICAL ANATOMY OF THE ZEBRAFISH RETINA

Distributors for North, Central and South America:
Kluwer Academic Publishers
101 Philip Drive
Assinippi Park
Norwell, Massachusetts 02061 USA
Telephone (781) 871-6600
Fax (781) 681-9045
E-Mail < kluwer@wkap.com>

Distributors for all other countries:
Kluwer Academic Publishers Group
Post Office Box 322
3300 AH Dordrecht, THE NETHERLANDS
Telephone 31 786 576 000
Fax 31 786 576 474
E-Mail < services@wkap.nl>

Electronic Services < http://www.wkap.nl>

Library of Congress Cataloging-in-Publication Data

Chemical anatomy of the zebrafish retina / by Stephen Yazulla ... (et al.).
p. cm.
"A special issue of Journal of neurocytology, a journal of cellular neurobiology, volume 30:7 (2001)"
ISBN 1-40207-086-1 (alk. paper)
1. Neurochemistry. 2. Immunocytochemistry. 3. Retina. 4. Zebra danios as laboratory animals. I. Yazulla, Stephen. II. Journal of neurocytology.

QP356.3 .C44 2002
573.8'8417482—dc21 2002141594

Printed on acid-free paper.
Printed in the United States of America

Journal of Neurocytology **30**, 549 (2001)

General introduction

The rate of growth of information generated by the application of molecular genetic techniques to studies of the nervous system has been astounding. From older studies on *Drosophila* to the Human Genome Project, the potential to unravel the secrets of neural function has never been greater. Currently, two important animal models, amenable to transgenic approaches, are being used with great success to study vertebrate neural function: the mouse and zebrafish. The retina, a part of the forebrain (diencephalon), is of particular interest because of its accessibility, our ability to control the external stimulus (light) and the opportunity to relate neurobiological data to the visual behavior of the organism. The retina is a complex tissue composed of six major types of neuron, each containing numerous anatomical and neurochemical subtypes, resulting in a huge diversity of neurons. The number of neuroactive substances found in the retina is also challenging, including amino acids, biogenic amines, purines, numerous neuropeptides, eicosanoids and acetylcholine. The retina extracts visual information from a narrow band of the electromagnetic spectrum we term "visible" light, though that band is subtly different for various animals, such as the zebrafish. From these spectra, the retina encodes intensity, contrast, wavelength (color), spatial frequency, position, perhaps movement, etc. over a 13 log-unit operating range of intensity. As the retina has a dynamic range of but 3 log units, powerful neural mechanisms reposition the dynamic range along this large operating range. Such adaptive control is further complicated by circadian mechanisms that modulate retinal sensitivity and metabolism. Spatial resolution of any visual pathway is limited by the retinal grain, the size of the cellular subunit underlying the function, and the retina is composed of many types of repeating subunits arranged in sheets across the retinal surface, some tiled, some overlapping. In addition, neuron types with dendrites extending for large distances range far outside the subunit structure, providing for extensive lateral communication. The key problem with any neural network, particularly with one as complicated as the retina, is assigning functions to a specific neuron type or subunit assembly. Genetic manipulation to isolate neuron types is one approach to this problem, but requires deep knowledge of the cellular organization of wild type or normal animals, a topic that this Special Issue is designed to address.

The zebrafish has become useful for genetic research because of its rapid development, fast generation time and the relative ease with which mutations may be produced and screened for developmental and behavioral defects. And the zebrafish is an explicitly visual model with proven tractability in rapid screening of high-level visual functions, unlike the largely nocturnal, visually impoverished mouse and it allies. The purpose of this Special Issue is to present a systematic description of the neurochemical anatomy of the zebrafish retina using light microscopic immunocytochemistry. The first paper tabulates all of the types and many of the specific neuronal subtypes in zebrafish retina by their intrinsic small molecule signatures, especially those with glutamatergic, GABAergic and glycinergic phenotypes, with attention to both mature retina and the marginal zone. The second paper presents an extensive survey of major transmitter systems including their biosynthetic enzymes, receptors and transporters as well as a representative sample of voltage-gated channels and membrane proteins. These independent efforts strongly validate the zebrafish as a typical albeit highly specialized cyprinid, enabling use of vast repository of prior electrophysiological, pharmacological and structural data gleaned from numerous cyprinid fishes. Together, we hope that these papers provide a foundation for design and analysis of experimental and genetic manipulations of the neuronal architecture the zebrafish retina.

Stephen Yazulla
Robert Marc

Journal of Neurocytology **30**, 551–592 (2001)

Neurochemical anatomy of the zebrafish retina as determined by immunocytochemistry

STEPHEN YAZULLA* and KEITH M. STUDHOLME

Department of Neurobiology and Behavior, Stony Brook University, Stony Brook, NY 11794-5230, USA
Stephen.Yazulla@Stonybrook.edu

Received 7 November 2001; revised 25 January 2002; accepted 25 February 2002

Abstract

The zebrafish retina is rapidly becoming a major preparation for the study of molecular genetic mechanisms underlying neural development and visual behavior. Studies utilizing retinal mutants would benefit by the availability of a data base on the distribution of neurotransmitter systems in the wild-type fish. To this end, the neurochemical anatomy of the zebrafish retina was surveyed by light microscopic immunocytochemistry. An extensive series of 60 separate antibodies were used to describe the distribution of major transmitter systems and a variety of neuron-associated membrane channels and proteins. These include markers (*i.e.*, antibodies against enzymes, receptors, transporters) for transmitters: GABA, glycine, glutamate, biogenic amines, acetylcholine, cannabinoids and neuropeptides; as well as a sample of voltage-gated channels and synapse associated membrane proteins. Discussion of the comparative localization of these antibodies is restricted to other teleost fishes, particularly goldfish. Overall, there was great similarity in the distribution of the various markers, as might be expected. However, there were some notable differences, including several antibodies that did not label zebrafish at all, even though goldfish retinas that were processed in parallel, labeled beautifully. This survey is extensive, but not exhaustive, and hopefully will serve as a valuable resource for future studies of the zebrafish retina.

Introduction

The rate of growth of information generated by the application of molecular genetic techniques to studies of the nervous system has been astounding. From the older studies on *Drosophila* to the Human Genome Project, the potential to unravel the secrets of neural function has never been greater. Currently, two important systems that are amenable to transgenic approaches are being used with great success to study vertebrate neural function: the mouse and zebrafish (Brockerhoff, 2001, for review). The retina, a part of the forebrain (diencephalon), is of particular interest to many neurobiologists because of its accessibility, the ability to control the external stimulus (light) and the opportunity to relate neurobiological data to the visual behavior of the organism. The retina is a complex tissue, comprising six major types of neuron: photoreceptors, horizontal cells, bipolar cells, amacrine cells, interplexiform cells and ganglion cells, each of these contains numerous subtypes. For example, the human retina has three types of horizontal cell and twenty-four types of amacrine cell (Kolb *et al.*, 1992). Teleost fish may contain even more subtypes, *i.e*, forty-three basic subtypes of amacrine cell in the roach that increases to seventy subtypes if differences in lamination are included as distinguishing criteria (Wagner & Wagner, 1988). The point is that given this huge diversity of neuron types, it is extremely difficult to assign a particular role in visual coding to any subtype of neuron, except for the obvious role of the photoreceptors. This task is easier if the cell type can be deleted selectively without collateral damage to other cell types. This was the case for dopamine interplexiform cells in fish that, following destruction with 6-hydroxydopamine, were found to be critical for the normal transition from dark to light adaptation (Lin & Yazulla, 1994a; McCormack & McDonnell, 1994; Li & Dowling, 2000). More recently, Yoshida *et al.* (2001) used immuno-toxin cell targeting techniques and showed that starburst amacrine cells were a key element in the circuit for directional sensitivity of ganglion cells and optokinetic nystagmus in mice. Another powerful technique is the use of knockout models that can be used to delete specific molecules in addition to cell types. For example, the role of the OFF pathway can be studied in animals that have had the ON pathway largely deactivated by knocking out the gene encoding the mGluR6 metabotropic glutamate

* To whom correspondence should be addressed.

0300–4864

receptor (Masu *et al.*, 1995). Complications with this technique are the compensatory changes that may take place among residual systems during development. Regardless of the technique used, the successful interpretation of these studies requires a detailed description of the cellular elements in the retina under consideration. Although Golgi studies provide excellent data on the structure of neuronal types, no information about the spatial organization, frequency or cell biology of the neuron is provided. For these data, immunocytochemistry has been invaluable. Studies on the effects of mutations need as a foundation, the cellular organization of wild type or normal animals. A good start for providing these types of data has appeared for the mouse retina (Yazulla *et al.*, 1997; Haverkamp & Wässle, 2000). As yet, a systematic description of the neurochemical anatomy of the zebrafish retina is not available.

The zebrafish has become very useful for genetic research because of its rapid development, generation time and relative ease with which mutations may be produced and screened for developmental and behavioral defects. There has been a rapid growth in papers on the zebrafish retina and visual system (Baier, 2000; Malicki, 2000; Bilota, 2001, for reviews). For example, a Medline search of "retina" and "zebrafish" revealed only 10 papers between 1990 and 1994, but 112 papers in the three years from June 1998 to August 2001. For the most part these papers deal with development in many aspects including descriptive studies, molecular genetics, growth factors, etc (Easter, 1996; Fadool, 2001, for reviews). There is an increasing effort to produce mass mutations and then screen the mutants for visual defects (*i.e.*, Brockerhoff *et al.*, 1995; Neuhauss *et al.*, 1999; Li, 2001). On-going and subsequent studies on the behavior, electrophysiology and anatomy of these mutants will lead to an understanding of the underlying bases of the visual defects. As with the mouse, a neurochemical atlas of the zebrafish retina would be of great use in future studies of zebrafish retina. Compared with other species there is relatively very little information regarding the zebrafish retina. The distributions of GABA, GAD, glycine, GluR5-7 and PKC have been described (Sandell *et al.*, 1994; Connaughton *et al.*, 1999; Caminos *et al.*, 2000; Nelson *et al.*, 2001) as well as glutamine synthetase and carbonic anhydrase (Peterson *et al.*, 2001). The purpose of this study was to provide an extensive description of the neurochemical anatomy of the zebrafish retina by light microscopic immunocytochemistry. The list of antibodies that we used is extensive but not exhaustive as we could not hope to localize all possible molecules of interest in the retina. Our goal was to provide a comprehensive start, localizing the major transmitter systems and a representative sample of voltage-gated channels and membrane proteins. We have not included any growth factors, second messengers (except PKC) or peptide receptors. Positive findings are reported for 60 antibodies; another 13 were tried but these did not stain the zebrafish retina.

Materials and methods

SUBJECTS

Zebrafish (*Brachydanio rerio*), 15 mm standard body length, were obtained from a commercial supplier and maintained at 23°C in aerated tanks filled with tap water circulating through an activated charcoal/filter system. The fish were fed Total Goldfish Flakes™ and kept on a 12 hour on/12 hour off light-dark cycle. Animals were treated according the guidelines of the National Institutes of Health and Association for Research in Vision and Ophthalmology.

PREPARATION

Light-adapted zebrafish were transected cervically at midday; dissections took place in room light. The cornea and lens were removed and the eyecups were fixed for 30 to 60 minutes at room temperature in 0.1 M phosphate-buffered 4% paraformaldehyde (pH 6.5). After rinsing in a 0.1M phosphate buffer (PB) pH 7.4, tissue was cryoprotected in 30% sucrose overnight. The eyecups were embedded in Tissue Tek O.C.T. (Ted Pella, Reading, CA) in an aluminum boat and frozen in liquid nitrogen-cooled isopentane. The 10–14 μm thick sections were mounted on gelatin/chromium coated slides, dried and stored in a non-frost-free freezer at −20°C. Sections were washed in phosphate buffered saline (PBS) pH 7.4, 10 minutes 3×, blocked in 2% normal goat serum (NGS) in PBS for 20 minutes. Sections were incubated overnight with primary antibody at 4°C in PBS containing 0.3% Triton X-100 and 5% NGS. After washing (15 minutes, 3X, in PBS) sections were incubated in the appropriate secondary antisera, rinsed, coverslipped with Vectashield (Vector Labs, Burlington, CA) and stored at −20°C. Sections were observed on a Zeiss Axioskop 2 epifluorescence microscope equipped with filter sets that were designed for FITC and Cy3 viewing. An additional FITC narrow-band pass filter (D535, Chroma Technology Corp. Brattleboro, VT) was inserted when viewing FITC to further insure no crossover from the Cy3. Sections, labeled for either FITC or Cy3 alone, showed no evidence of crossover fluorescence when viewed or photographed with the alternate filter set.

PRODUCTION OF PHOTOMICROGRAPHS

In general, figures of single-labels were obtained with a Zeiss 40 × dry planapochromatic objective (n.a., 0.75). Double-label figures were, in general, obtained with a Zeiss 63 × dry planapochromatic objective (n.a., 0.95). Light micrographs were acquired as TIFF files directly from the Zeiss microscope using a Kodak DCS 290 digital camera at 1792 × 1200 (pixels). All TIFF files were optimized for brightness and contrast using Adobe PhotoShop (v. 5.5). Final figures were composed and labeled using Corel Draw (v. 9).

ANTISERA

Primary antisera are listed in Table 1. Antibodies in bold did not label the zebrafish retina. All secondary antisera were

Table 1. List of antibodies used in the current study (bold—did not label zebrafish retina).

Antibody against	*Animal*	*1° dilution*	*Source*	*Figure*
AChR neuronal nicotinic (chicken brain)	mouse	1:20–1:50	Devel. Studies Hybrid. Bank, Univ Iowa, Iowa City, IO (mAb 270)	X
Adenosine deaminase (ADA; calf spleen)	rabbit	1:1,000	Chemicon International Inc., Temecula, CA (AB176)	Figure 17B
Calbindin-D 28k (clone CL-300)	mouse	1:500	Sigma Immunochemicals, St. Louis, MO (#C-8666)	X
Calretinin (guinea pig)	rabbit	1:1,000–1:4,000	Chemicon International Inc. (AB149)	Figure 3B
Calretinin (recombinant rat)	mouse	1:4000	Chemicon International Inc. (MAB1568)	not shown
Cannabinoid CB1 receptor (AA1-14 rat CB1R)	rabbit	1:1,000	H. McIntosh *et al.* (1998)	Figure 19A
Cholecystekinin (CCK)	rabbit	1:20,000	Incstar Corp, Stillwater, MN	Figure 20A
Cystic fibrosis trans regulator (CFTR; AA729-736)	mouse	1:50	Genzyme, Cambridge, MA (#1660-01))	Figure 25A
Choline acetyltransferase (ChAT; human placenta)	goat	1:2,000	Chemicon International Inc. (AB144P)	Figure 18
ChAT (porcine brain)	mouse	1:2,000–1:4,000	Boerhringer Mannheim, Indianapolis, IN (#770981)	not shown
Connexin 26 (CX26; rat CX26))	rabbit	1:100	Zymed Laboratories, San Francisco, CA (#71-0500)	Figure 25C
Dopamine beta hydroxylase (DBH; bovine adrenal)	rabbit	1:400	Eugene Tech International, Inc., Ridgefield Park, NJ (TE103)	eye, not retina
Dopamine D1 receptor (gfD1R; C-terminal gfD1R)	chicken	1:1,000	Mora-Ferrer *et al.* (1999)	Figure 16A
Enkephalin(leu)	rabbit	1:15,000	Immunotechnology, Inc., Chapel Hill, NC (#RE1)	Figure 21A
Enkephalin(met)	rabbit	1:500	R. Altschuler (NIH, Batch RA143)	Figure 21C
Fatty acid amide hydroxylase (AA 38-579 rat FAAH)	rabbit	1:250	B. Cravatt (Patrecelli *et al.*, 1998)	Figure 19B
FMRF-amide	rabbit	1:6,000	Cambridge Research Biochemical, Wilmington, DE (#CA245)	Figure 20B and C
$GABA_A\alpha1$(AA1-16 rat $GABA\alpha1$)	rabbit	1:200	Alomone Labs, Jerusalem, Israel (#AGA-001)	Figure 8A
$GABA_A\alpha3$ (AA1-15 human $GABA\alpha3$)	rabbit	1:300	Alomone Labs (#AGA-003)	Figure 8B and D
$GABA_A\beta2/\beta3$ (mAb 62-3G1; purified rat subunits)	mouse	1:5	A.L. deBlas (Vitorica *et al.*, 1988; Lin & Yazulla, 1994)	Figure 9A
$GABA_A\beta2/\beta3$ (mAb BD-17; purified rat subunits)	mouse	1:1	J.G. Richards (Schoch *et al.*, 1984; Richards *et al.*, 1987)	Figure 9B
$GABA_C$(ρ; N-terminal $\rho1$)	rabbit	1:100	R. Enz *et al.* (1996)	Figure 9C
GABA transaminase (GABA-T; mouse brain GABA-T)	rabbit	1:600	J.-Y. Wu (Lin *et al.*, 1983)	Figure 6
Glutamic acid decarboxylase (GAD; catfish brain GAD)	rabbit	1:600	J.-Y. Wu (Su *et al.*, 1979; Yazulla *et al.*, 1986)	Figure 5C
GAD (1440-4; rat brain GAD)	sheep	1:750	W. Oertel *et al.* (1981)	Figure 5A
GAD 65 (GAD6; rat brain)	mouse	1:100	Developmental Studies Hybridoma Bank	X
GAD 67 (K2; feline brain)	rabbit	1:800, 1:1,200	Chemicon International Inc. (AB108)	Figure 5B and D
GABA transporter (GAT1; C-terminal rat GAT1)	rabbit	1:15,000	N. Brecha (Johnson *et al.*, 1996)	Figure 7a
GAT 2 (C-terminal rat GAT2)	rabbit	1:800	N. Brecha (Johnson *et al.*, 1996)	Figure 7B
GAT 3 (C-terminal rat GAT3)	rabbit	1:1,000	N. Brecha (Johnson *et al.*, 1996)	Figure 7C
GABA vesicular transporter (C-terminal rat VGAT)	rabbit	1:1,000	Chemicon International Inc. (AB5062P)	Figure 7D
Gephyrin (mAb7A; rat spinal cord)	mouse	1:125	Boerhringer Mannheim	Figure 10B and C
Glial fibillary acidic protein (GFAP; pig spinal cord)	mouse	1:500	Chemicon International, Inc. (MAB360)	Figure 26

(*Continued on next page.*)

Table 1. (*Continued*).

Antibody against	*Animal*	*1° dilution*	*Source*	*Figure*
Glucagon	rabbit	1:400	Arnel Products Co. Inc., Cherokee Station, NY	X
Glutamate receptor1 (C-terminal rat GluR1)	rabbit	1:10	Chemicon International Inc. (AB1504)	Figure 11A
Glutamate receptor2 (AA175-430 fusion protein)	mouse	1:75	Chemicon International Inc. (MAB397)	Figure 11B
Glutamate receptor2/3 (C-terminal rat GluR2)	rabbit	1:400	Chemicon International Inc. (AB1506)	Figure 11C
Glutamate receptor4 (C-terminal rat GluR4)	rabbit	1:10	Chemicon International Inc. (AB1508)	Figure 11D
mGluR1α (C-terminal rat mGluR1α)	rabbit	1:10	Chemicon International Inc. (AB1551)	Figure 13A
mGluR2/3 (C-terminal rat mGluR2)	rabbit	1:10	Chemicon International Inc. (AB1553)	Figure 13B
mGluR5 (C-terminal rat mGluR5)	rabbit	1:750	Chemicon International Inc. (AB1596)	Figure 13D
Glutamate transporter (C-terminal rat GLT-1)	guinea pig	1:100	Chemicon International Inc. (AB1783)	Figure 14
Glycine Receptor α1 (MAb2b)	mouse	1:50	H. Betz (Schroder *et al.*, 1991)	Figure 10A
Kainate receptor (GluR5,6,7; recombinant GluR5)	mouse	1:3,000–4,000	Chemicon International Inc. (MAB379)	Figure 12A
Kainate receptor (GluR6,7; C-terminal rat GluR6)	rabbit	1:500	Upstate Biotechnologies Inc., Lake Placid, NY (#06-309)	Figure 12B
Kainate receptor (KA2; C-terminal KA2)	rabbit	1:750	Upstate Biotechnologies Inc. (#06-315)	Figure 12C
Neurotensin	rabbit	1:100–1:1,000	Peninsula Labs Iinc., Belmont, CA	X
Neuropeptide Y (NPY)	rabbit	1:2,000	Incstar Corp,	Figure 22A
NMDA receptor NR1 (C-terminal rat NR1)	rabbit	1:600	Chemicon International Inc. (AB1516)	X
NMDA receptor NR2b (AA984-1104 fusion protein)	rabbit	1:100	Chemicon International Inc. (AB1557P)	Figure 12D
Parvalbumin (carp muscle; clone PA235)	mouse	1:1,000	Sigma (#P3171)	X
Phenylethanolamine N-methyltransferase (rat PNMT)	rabbit	1:1K	M. Bohn *et al.* (1987)	Figure 15B
Postsynaptic density 95 (PSD-95)	mouse	1:5	J. Trimmer	Figure 25B
Potassium channel (Kv1.1; AA416-495 mouse Kv1.1)	rabbit	1:75	Alomone Laboratores (#APC-009)	Figure 24A
Potassium channel (Kv1.2; AA417-499 rat Kv1.2)	rabbit	1:500	Alomone Laboratores (#APC-010)	Figure 24B
Potassium channel (Kv1.4; AA13-37 *Shaker* Kv1.4)	mouse	1:10	J. Trimmer (Bekele-Arcuri *et al.*, 1996)	Figure 24C
Potass. chan (Kv2.1; AA837-853 rat Kv2.1)	rabbit	1:25	J. Trimmer	X
Potassium channel (Kv4.2; C-terminal rat Kv4.2)	rabbit	1:25	J. Trimmer (Sheng *et al.*, 1992)	Figure 24D
Protein Kinase C α/β(PKC; MC5 clone; bovine brain)	mouse	1:75, 1:200	Amersham (#RPN.536)	Figure 3A, 8C, 13C
Recoverin	rabbit	1:25K	K.-W. Koch (Lambrecht and Koch, 1991)	Figure 3C, 4B
Sodium Channel—Pan (peptide in III-IV linker region)	rabbit	1:400	S.R. Levinson (Dugandzija-Mivakovic *et al.*, 1995)	Figure 23A
Sodium Channel—(Na 1.2 subunit C-terminus)	rabbit	1:500	J. Trimmer (Gong *et al.*, 1999)	Figure 23C
Synaptic associated protein-97 (SAP-97)	rabbit	1:1,000	Affinity BioReagents, Neshanic Station, NJ	X
Serotonin (5HT; clone YC5/45)	rat	1:2,000	Sera-Labs Limited, Sussex, UK (#MAS 055)	Figure 17A
Somatostatin (SRIF; SOMA8)	mouse	1:1,000	A. J. Malcom	Figure 22C and D

(*Continued on next page.*)

Table 1. (*Continued*).

Antibody against	*Animal*	*1° dilution*	*Source*	*Figure*
Substance P (C-terminal, clone NC1/34 HL)	rat	1:1,000	Pelfreeze Biologicals, Rogers, AZ (#MAS 035b)	Figure 22B
Synaptic vesicle 2 (SV2; ommata elctric organ)	mouse	1:2000	Developmental Studies Hybridoma Bank	Figure 25D
Tyrosine hydroxylase (TOH; bovine adrenal)	rabbit	1:800	Eugene Tech International (TE101)	Figure 15A
Tyrosine hydroxylase (TOH; quail)	mouse	1:75	Developmental Studies Hybridoma Bank	X
Vasoactive Intestinal Peptide (VIP)	rabbit	1:100–1:1,000	Incstar	X
Vimentin—14H7	mouse	1:100	Developmental Studies Hybridoma Bank	X
Zns-5	mouse	1:500	Zebrafish Monoclonal Antibody (MAB) Facility, Eugene OR	X
Zpr-1 (FRet43)	mouse	1:750	Zebrafish MAB Facility (Larison & BreMiller, 1990)	Figure 2A, 4A
Zpr-2 (FRet45)	mouse	1:750	Zebrafish MAB Facility	Figure 2B
Zns-2	mouse	1:1,000	Zebrafish MAB Facility	Figure 2C

made in donkey conjugated to FITC or Cy3 and obtained from Jackson ImmunoResearch Labs (West Grove, PA).

Results

It is important to note that most of the antisera used in this study have not been characterized in the zebrafish retina. What is described here is the pattern that was obtained with a particular antibody. Although it is likely that many of the antibodies recognized the appropriate antigen in this tissue, we are not claiming any functional implications of these patterns. Each section of the Results will contain the observations and a discussion of the comparative distribution of the antibody in other fishes rather than an exhaustive comparison with all species. We have attempted to be as comprehensive as possible when citing earlier work on the fish retina; we apologize in advance for inadvertent omissions.

One of the more surprising findings was the number of differences in the labeling patterns of antibodies between zebrafish and goldfish. In addition, several antibodies produced no labeling whatsoever in zebrafish even though goldfish retinas that were processed identically side-by-side with the zebrafish showed excellent labeling.

The overall scheme for the presentation of the data is based on a particular transmitter system, *i.e.*, GABA, including biosynthetic enzymes, transporters and receptors as opposed to grouping by function, *i.e.*, receptors, regardless of the transmitter system. There are other sections on neuropeptides, ion channels, calcium binding proteins and miscellaneous topics. Many of the antibodies labeled small structures in the outer plexiform layer that could have included photoreceptor terminals, the dendrites of horizontal cells, bipolar cells or interplexiform cells, or processes of Müller's cells. A series of double-label plates are presented in a separate section at the end that attempt to identify these structures.

An overview of the zebrafish retina is illustrated in Figure 1. This is a Nomarski view of a cryostat section and it serves as a reference for the fluorescence micrographs to follow. This section and all to follow was obtained from a light-adapted retina, identifiable by the retracted cones and presence of retinal pigment epithelium (RPE) granules around the cone outer segments. A conventional nomenclature will be used. "Proximal" means towards the vitreal surface and "distal" towards the retinal pigment epithelium. Lamination in the inner plexiform layer (IPL) will be designated as layers 1–5 of equal thickness, with layer 1 the most distal and layer 5 the most proximal. The distal portion of the IPL, comprising layers 1–3, will be referred to as sublamina *a*, and is associated with OFF responses to light. The proximal portion, comprising layers 4 and 5, will be referred to as sublamina *b*, and is associated with ON responses to light (Famiglietti *et al.*, 1977; Stell *et al.*, 1977). The use of "mid" will refer to the middle (*i.e.*, 50% position) of any given layer.

Photoreceptors occupy about one-third of the retinal thickness, and consist of rods and cones, with the larger cone cell bodies distal to the rods in the outer nuclear layer (ONL). Cell bodies of horizontal cells, bipolar cells and amacrine cells occupy the distal, mid and proximal regions of the inner nuclear layer (INL), respectively. Cell bodies in the ganglion cell layer (GCL) include ganglion cells and displaced amacrine cells.

Immuno-labeling at the outer limiting membrane (OLM), inner segments and nuclei of the photoreceptors was observed for many antibodies, not only here for zebrafish but in other species as well (*i.e.*, Enz *et al.*, 1996; Vitanova *et al.*, 2001). Therefore, unless otherwise stated, the episodic intense labeling of photoreceptors will not be mentioned for each antibody; we

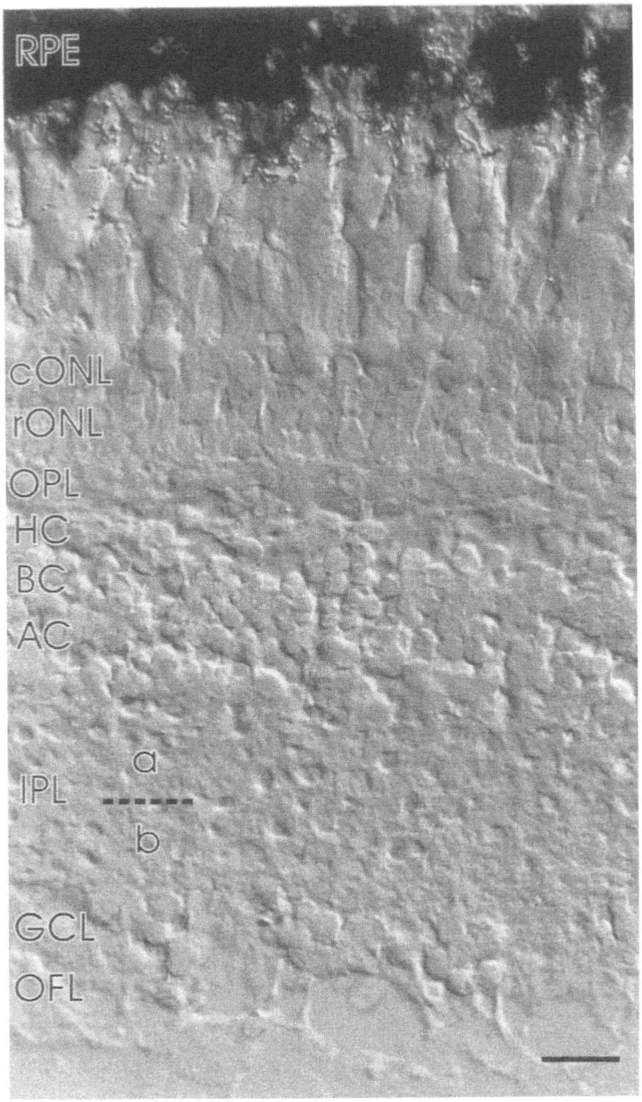

Fig. 1. Light micrograph of the zebrafish retina obtained from a 15 μm cryostat section and viewed with Nomarski DIC optics. RPE—retinal pigment epithelium, cONL—cone outer nuclear layer, rONL—rod outer nuclear layer, OPL—outer plexiform layer, HC—horizontal cells, BC—bipolar cells, AC—amacrine cells, INL—inner nuclear layer, IPL—inner plexiform layer is divided into a distal sublamina a and a proximal sublamina b. GCL—ganglion cell layer, OFL—optic fiber layer. Scale bar = 20 μm.

only indicate its presence here without attributing any importance to it.

ZEBRAFISH SPECIFIC ANTIBODIES

We obtained four zebrafish-specific antibodies from the Zebrafish Hybridoma Facility at the University of Oregon that seemed to have promise for the retina.

The first and most impressive antibody, ZPR1 (a.k.a. Fret-43) has been reported to label double cones in zebrafish (Larison & BreMiller, 1990). We show the pattern here for completeness and to set up further double-label experiments to verify localization within the OPL (Fig. 2A). ZPR1-IR was completely restricted to double cones, labeling them from the outer segment to the synaptic terminal. The cells were beautifully labeled including the axon. The inner segment including the paraboloid was only outlined, and appeared as a dark oval between the nucleus and outer segment. ZPR1-IR labeled double cones as well as some single cones in the carp and goldfish retinas (Schultz *et al.*, 2001; Yazulla & Studholme, unpublished observations), and as such is not an unequivocal marker for double cones in carp and

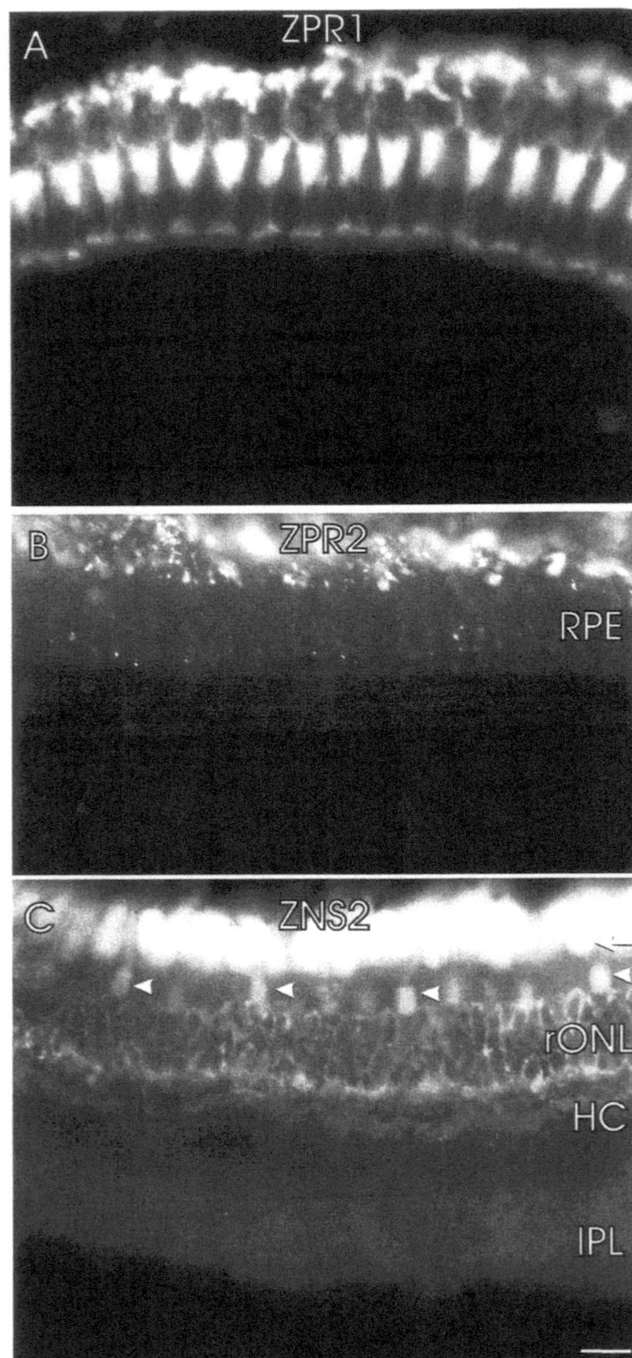

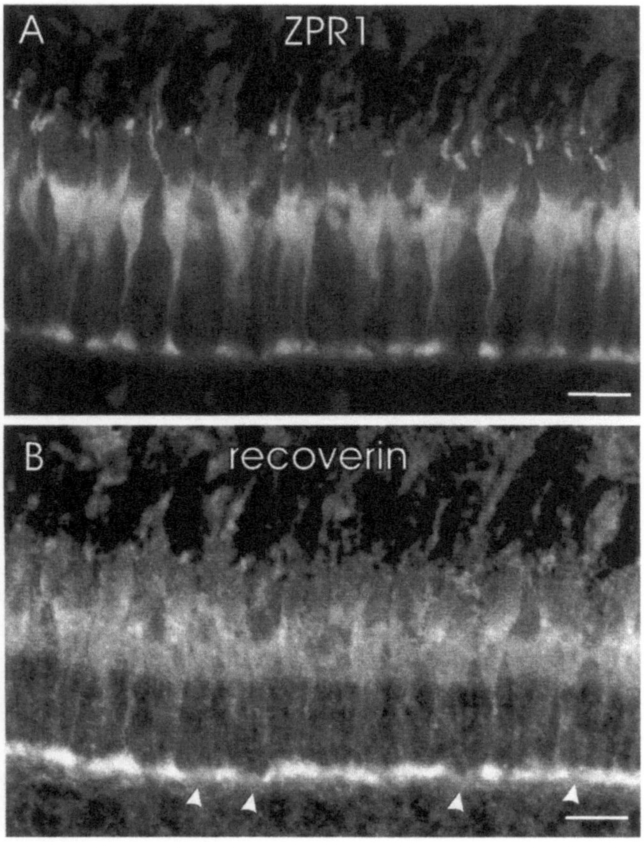

Fig. 3. ZPR1/recoverin double label shows that all ZPR1-IR (A) cones were recoverin-IR (B). However, there are some gaps in the ZPR1-IR pattern that contain recoverin-IR (arrowheads) indicating that ZPR1-IR stains the vast majority but not all cones. Scale bar = 20 μm.

goldfish, as it appears to be in zebrafish (see Fig. 3 for ZPR1/recoverin double label).

ZPR2-IR labeled only the retinal pigment epithelium, including the fine processes that extended proximally between the cone inner segments (Fig. 2B) although these processes were labeled only weakly. ZNS2-IR (Fig. 2C) was most prominent over the entire photoreceptor region, weak over horizontal cells and weaker yet over the IPL. It appeared as if all cone inner segments were labeled including the double cones (horizontal arrow) and short single cones (arrowheads). Note the difference with ZPR1-IR in which there was no label over the inner segments. The photoreceptor labeling was interesting in that rod nuclei were outlined by label whereas it was the cone inner segments that were filled with label. It is unlikely that the label surrounding the rod nuclei was derived from cone processes traversing the ONL because this was not observed with ZPR1-IR, which shows cones axons projecting directly to the OPL. The hint of a band at the outer limiting membrane suggests that the ZNS2-IR in the rod ONL was in Müller's cell processes. ZNS5 did not label the zebrafish retina.

Fig. 2. Distribution of zebrafish-specific antibodies. (A) ZPR1-IR labels double cones from the outer segment to the synaptic pedicle. The paired axons extending from the nucleus to the pedicle are easily visible. (B) ZPR2-IR was restricted to the retinal pigment epithelium (RPE); it was densest at the apical end and was less in processes that extended proximally between the cones. (C) ZNS2-IR appeared to stain all cones intensely as well as light staining of a layer of horizontal cell bodies (HC) and the inner plexiform layer (IPL). The staining differed from ZPR1 in that the inner segments, rather than the nuclei, were stained. Also, the inner segments of short single cones (arrowheads) as well as the double cones (horizontal arrow) were ZNS2-IR. The rod nuclei (rONL) were outlined with ZNS2-IR. It is not clear whether this staining was in the rods or Müller's cell processes that surrounded the rod nuclei. Scale bar = 20 μm.

PROTEIN KINASE C AND CALCIUM BINDING PROTEINS

We were unable to detect any labeling with either parvalbumin or calbindin antibodies, consistent with an earlier report that also was unable to detect parvalbumin and calbindin immunoreactivities in the goldfish retina (Hamano *et al.*, 1990).

Protein kinase C

PKC-IR was found in numerous cell bodies at about the 50% to 70% depth in the INL (Fig. 4A). The more distal group were fewer in number, had thicker dendrites that ascended to the OPL and thicker axons that descended to the IPL in which they ended as a relatively large synaptic terminal in the proximal IPL. The more numerous cell bodies had thinner dendrites and axons that appeared to terminate in small boutons in a layer that was more distal to the larger terminals. Caminos *et al.* (2000) only described the large PKC-IR bipolar cells in zebrafish, tench and goldfish. Of interest is that, unlike goldfish and tench, there was no regional variation in the size of the soma or axon terminal of the large PKC-IR bipolar cells in the zebrafish. The appearance of two groups of PKC-IR bipolar cells shown here in zebrafish is very similar to that reported by Suzuki and Kaneko (1990) in goldfish. Based on a comparison of position and size, it appears as if PKC-IR labels all ON bipolar cells in the zebrafish retina; the larger cells would correspond to the mixed rod/cone (M*b*) bipolar cells and the others by exclusion would be the cone (C*b*) bipolar cells. This became a useful tool when trying to differentiate ON bipolar cells from OFF bipolar cells as in the case of the PanNa-channel antibody. A difference between zebrafish and goldfish is that the C*b* bipolar cells were more brightly labeled by PKC-IR and they made up a higher proportion of ON bipolar cells in zebrafish. Particularly, in goldfish, the C*b* bipolar cells were very weakly labeled compared to the M*b* bipolar cells, and in order to detect them at all, often the

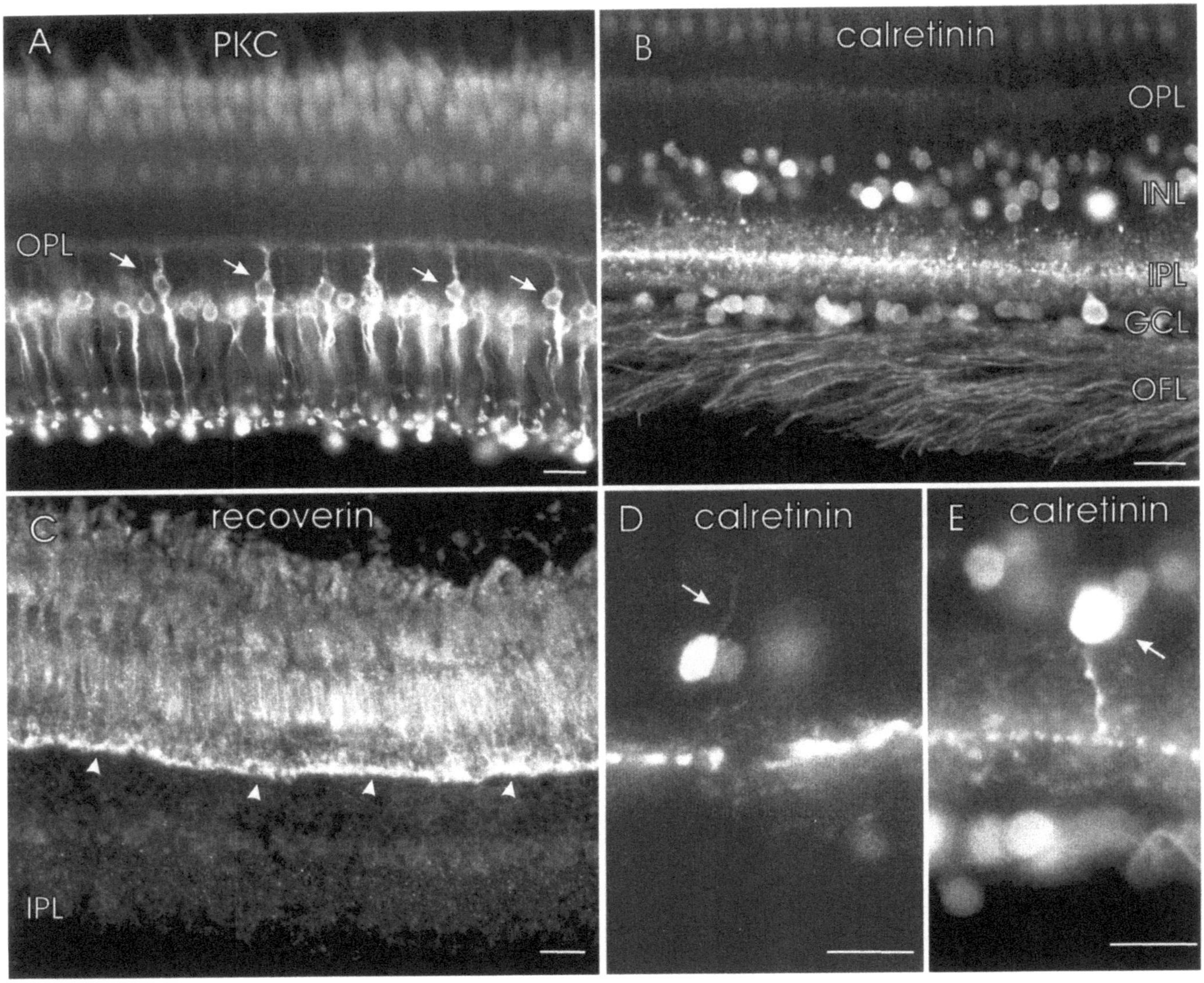

Fig. 4. Protein kinase C and calcium-binding proteins. (A) PKC-IR was restricted to bipolar cells that appeared to be of two general types. The first, less common, had a larger soma that was more distal in the INL (arrows), dendrites projecting to the OPL and a large synaptic terminal in layer 5 of the IPL. The second, more numerous type, had smaller cell bodies lined up in the proximal third of the INL with projections to small synaptic terminals in layer 4 of the IPL. Dendrites going to the OPL were faint. (C) Recoverin-IR was restricted to the photoreceptors that appeared to be diffusely stained. However, there was more intense staining over the synaptic pedicles (white arrowheads). (B, D and E) Calretinin-IR was broadly distributed in the retina but was most intense in a small population of amacrine cells that were unistratified in sublamina 3 (E). Numerous cell bodies in the mid INL (likely to be bipolar cells) and ganglion cell layer (GCL) including optic nerve fibers were labeled (B). The entire IPL was labeled with bouton-like structures rather than continuous fibers; these were concentrated in sublamina 3. Fine processes were observed projecting from cell bodies in the INL to the OPL (D, arrow). In the absence of descending processes, it is not clear if these cells were bipolar cells or a type of interplexiform cell. Scale bar = 20 μm (A, B and C), 10 μm (D and E).

goldfish tissue needed to be over-reacted *vis a vis* the M*b* bipolar cells.

Recoverin

Recoverin-IR was restricted to cone photoreceptors (Fig. 4C), labeling the entire cell from the terminal to outer segment. The synaptic terminal was more intensely labeled than the rest of the cell. The nuclei of rods appear to be outlined by recoverin-IR, rather than filled by it. There did not appear to be labeling in the region of rod spherules that would appear as small boutons just distal (above in micrograph) to the cone pedicles. Recoverin-IR was useful in combination with other antibodies to differentiate labeling in cone photoreceptor terminals from dendrites of horizontal cells or bipolar cells that invaginated the pedicle.

Calretinin

Calretinin-IR produced a very interesting and beautiful pattern in the retina (Fig. 4B,D, and E). There were numerous cell bodies in the proximal half of the INL and in the ganglion cell layer. It is likely that ganglion cell bodies as well as displaced amacrine cells were labeled because there was extensive calretinin-IR in the optic fiber layer and there were far more cell bodies in the ganglion cell layer than could be accounted for by ganglion cells alone. There also was a very dense band of label in layer 3 of the IPL. This band appeared to be derived from a type of pyriform amacrine cell (Fig. 4E). There was lighter, but significant, label in sublaminas *a* and *b*. The ovoid cell bodies in the mid INL often had processes that extended to the OPL (Fig. 4D). In the absence of descending processes, we are not sure if these cells were bipolar cells or the Type II glycine interplexiform cell described by Marc and Lam (1981) in goldfish.

GABAergic SYSTEM

The distributions of GABA, GAD-1440 and GAD-67 have been described previously in zebrafish retina (Sandell *et al.*, 1994; Connaughton *et al.*, 1999). Our findings are in agreement with these earlier papers and here we extend the data to include two additional antibodies against GAD, GABA-T, GABA receptors and GABA transporters.

Glutamate decarboxylase

Four GAD antisera were used, all of which robustly label the goldfish retina. Only three of the four labeled the zebrafish retina; mouse anti-GAD-65 produced no labeling at all. Overall, the three GAD antisera produced similar patterns with some differences (Fig. 5). The most common feature was labeling throughout the inner plexiform layer (IPL) with a dark band at about layer 3 of the IPL. Subtle lamination can be seen in the distal and proximal halves of the IPL as well. Distinct labeling of numerous amacrine cell bodies was most

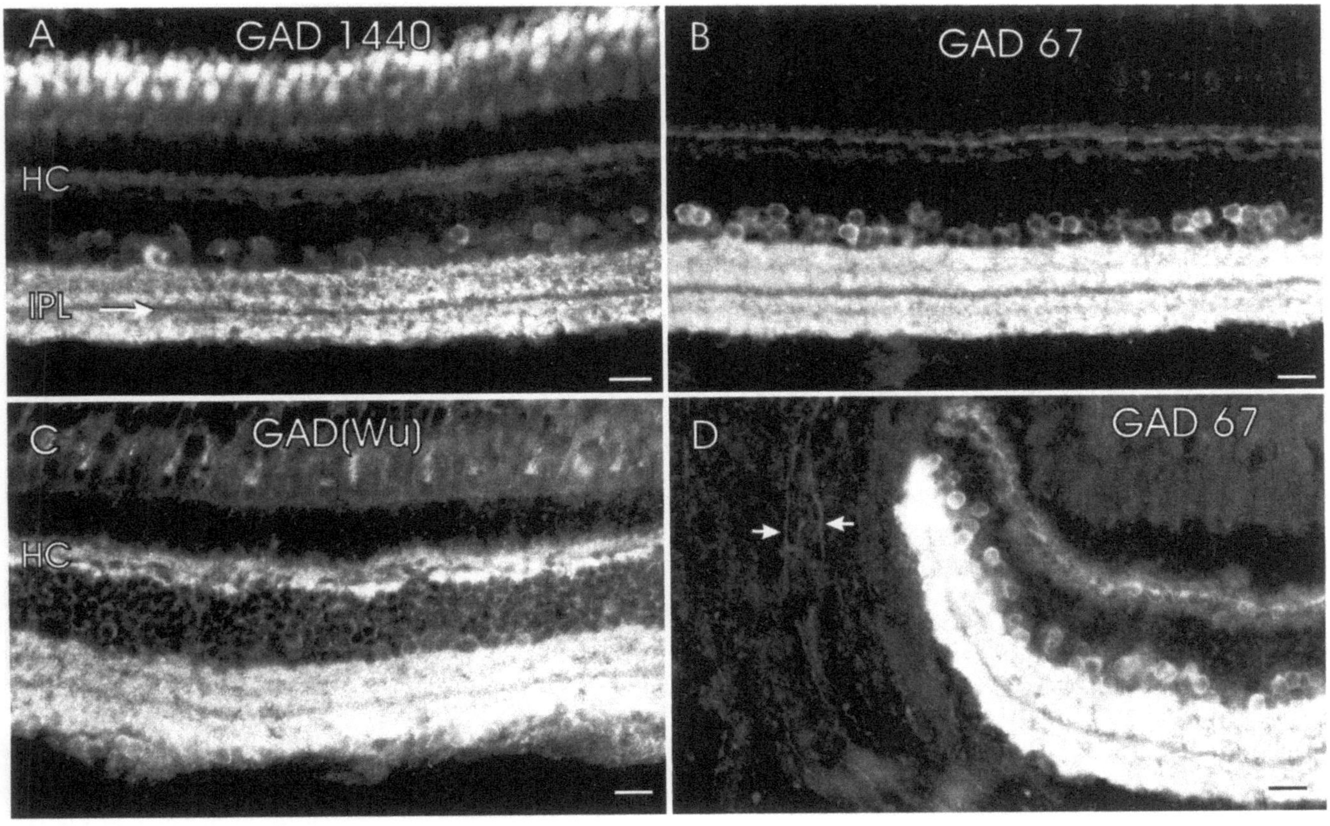

Fig. 5. Distribution of glutamic acid decarboxylase (GAD); Scale bars = 20 μm. (A) GAD-1440 intensely labeled the full thickness of the IPL, with a prominent dark band, just proximal to sublamina 3 (arrow). Lighter label was seen over numerous amacrine cells and the horizontal cell bodies (HC) in the distal INL. (B and D) GAD-67 produced a pattern nearly identical to GAD-1440, except that amacrine cells were labeled more intensely. Although cells in the ganglion cell layer were not labeled, GAD-67 was observed in optic fibers in the optic nerve head (D, arrows). (C) GAD-Wu from catfish brain GAD produced an identical pattern in the IPL as the other two GAD antisera, but amacrine cells were labeled less, whereas horizontal cells, including the OPL, were labeled more.

apparent with GAD-67, less so with GAD-1440 and least with anti-catfish brain GAD. Conversely, labeling of horizontal cell bodies and processes in the outer plexiform layer (OPL) was most prominent for anti-catfish brain GAD, slightly less so for GAD-67 and least for GAD-1440. Labeled cell bodies in the ganglion cell layer were most apparent for anti-catfish brain GAD and not detectable with GAD-1440. Sections though the optic nerve head revealed optic nerve fibers that were labeled for GAD-67. Note also the persistence of horizontal cell labeling right up to the border of the optic nerve head, in contrast to the report of Sandell *et al.* (1994) who showed that GABA-IR in horizontal cells was weak in the central retina, but more intense in peripheral retina.

The GAD-65 antibody (GAD6 of Chang & Gottlieb, 1988) did not label the zebrafish retina, although in goldfish we found that it produces a near identical but weaker pattern as GAD-67 in goldfish. It should be noted that the GAD-65 pattern reported in zebrafish by Connaughton *et al.* (1999) actually was obtained with the GAD-1440 antisera (Oertel *et al.*, 1981). GAD-1440 has broad recognition of GAD-65 and GAD-67 with a preference for GAD-65 (Kaufman *et al.*, 1991). However, the absence of specific GAD-65 (*i.e.*, GAD6) labeling in this study suggests that the GAD-1440 pattern is due to its recognition of GAD-67. This suggests that GAD-65, as recognized by the GAD6 antibody, is not present in the zebrafish retina. This may not be the last word because Martin *et al.* (1998) reported that zebrafish contain genes for both GAD-65 and GAD-67 that code for products with very high sequence homology to human GAD-65 and GAD-67. Although both GAD-65 and GAD-67 were found in many regions of the zebrafish brain, only GAD-67 was found in the caudal hindbrain. Unfortunately, the retina was not included in this list of regional localization, as it too, may contain only GAD-67.

This overall pattern in the zebrafish is similar to that described for GAD-IR in the goldfish using anti-catfish brain GAD (Zucker *et al.*, 1984; Yazulla *et al.*, 1986) and anti-rabbit brain GAD (Brandon, 1985), with two exceptions. First, there was no evidence of GAD-IR or GABA-IR (Connaughton *et al.*, 1999; Marc & Cameron, 2001) in the axon terminals of horizontal cells in the zebrafish, which are very thin structures unlike the larger "nematode-like" processes of goldfish. The axon terminals in goldfish meander through the mid to proximal INL and are labeled intensely by GAD-IR, GABA-IR and high affinity ^{3}H-GABA uptake (Yazulla, 1986, for review). Second, in goldfish IPL, GAD-IR appears in three broad laminas in layers 1, 3 and 5, with two dark bands in layers 2 and 4. This is in striking contrast to zebrafish IPL that has more discrete lamination in layers 1, 2, 4 and 5 and a dark band in layer 3.

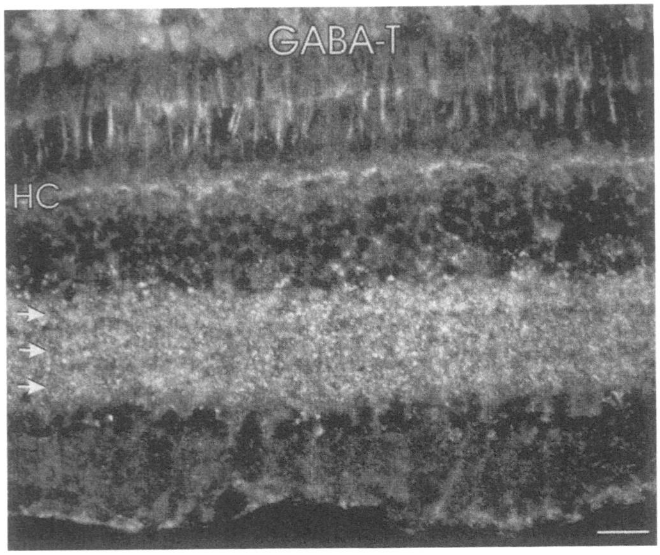

Fig. 6. GABA-T-IR appeared to be most intense in three broad laminas of the IPL (arrows), scattered amacrine cell bodies and surrounding horizontal cell bodies (HC), perhaps extending to dendrites in the OPL. Scale bar = 20 μm.

GABA-transaminase

The GABA metabolizing enzyme GABA-T was distributed through the retina, appearing rather punctate (Fig. 6). Label was found over horizontal cell bodies, scattered amacrine cell bodies and in three broad bands in the IPL. Curiously, there was no counterpart to the dark band in layer 3 as seen with anti-GAD. There was no labeling in the ganglion cell or optic fiber layers. There was however, a hint of vertical streaks through the optic fiber layer to the vitreal margin, indicative of the proximal region of Müller's cells. GABA-T-IR in zebrafish is very similar to that observed in goldfish (Yazulla, 1991) except that the three laminas in the IPL appear less well defined in zebrafish. The potential that GABA-T is present in Müller's cells is of interest because there is no evidence for the presence or uptake of GABA in Müller's cells of any non-mammal, except in the skate, an elasmobranch fish (Lam, 1975; Bruun *et al.*, 1984).

GABA transporters

We localized four GABA transporters: three that transport GABA across the plasma membrane and one that transports GABA into synaptic vesicles (Fig. 7). The plasma membrane GABA transporters GAT-1, GAT-2 and GAT-3 produced very different labeling patterns. Superficially, GAT-1 most resembled the GAD pattern with numerous amacrine cell bodies, a broad distribution in the IPL, more prominent in layers 1 and 4, and scattered cell bodies in the ganglion cell layer. There was only marginal GAT-1 labeling in the OPL. GAT-2

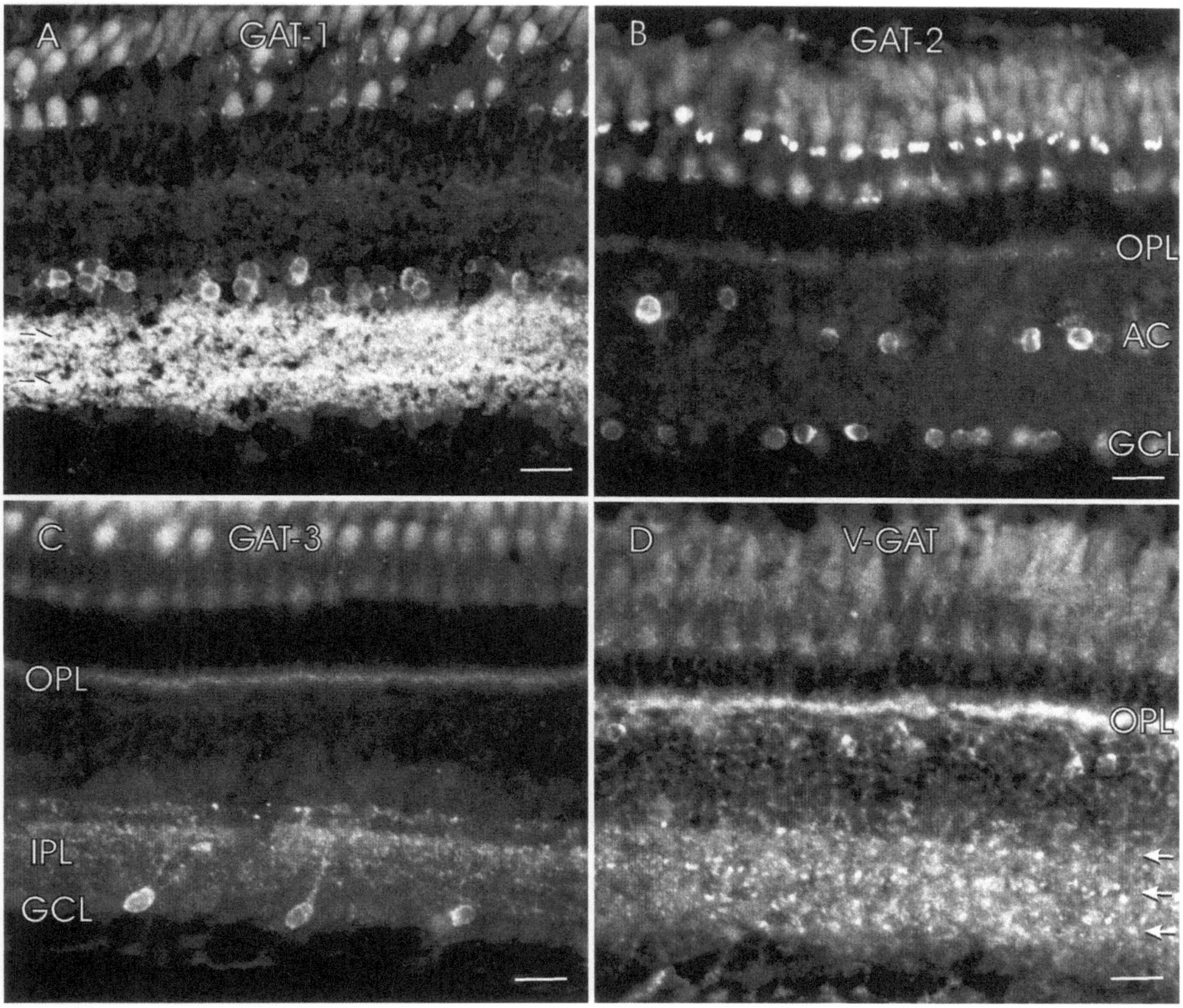

Fig. 7. GABA transporters. Scale bars = 20 μm. (A) GAT1-IR was most pronounced in amacrine cell bodies and the IPL, in which there appeared to be bands in layers 1 and 4 (arrows). (B) GAT2-IR was prominent in amacrine cell bodies (AC) and cell bodies in the ganglion cell layer (GCL). Weak label in the OPL also was noted. (C) GAT3-IR was found in cell bodies in the GCL that gave rise to dendrites that ramified in the distal layers of the IPL. GAT3-IR was concentrated in the distal IPL, with a broad band in layer 2 and a very narrow band in layer 1. There was also a narrow diffuse band of GAT3-IR in the OPL. (D) VGAT-IR was found in the distal INL and OPL, likely associated with horizontal cells, scattered cell bodies in the mid and proximal INL and in the IPL. GAT3-IR in the IPL was diffuse, but there appeared to be three layers of irregularly-spaced larger boutons in layers 1, 3 and 4/5 (arrowheads).

labeled numerous cell bodies at both the inner and outer margins of the IPL, which itself was not labeled. There were some very weakly labeled cell bodies in the mid to distal INL. Fine, discrete puncta were visible in the OPL. There was also a band of bright spots in the region of the base of the inner segments of double cones and long single cones, but not for short single cones. GAT-3 was unusual in that the only cell bodies labeled were in the ganglion cell layer and these often had dendrites projecting to both proximal and distal layers of the IPL. The distal half of the IPL was more intensely labeled for GAT-3; an isolated thin layer at the IPL/INL border was particularly noted. GAT-3 also labeled a thin band in the OPL in a region that appeared to include the cone terminals.

The vesicular GABA transporter, VGAT, was most intense in the OPL, in which it appeared as a narrow band at the level of horizontal cell dendrites (Fig. 7D). Label was also seen along the proximal margin of horizontal cell bodies indicating that VGAT labeled horizontal cells and their processes. Numerous cell bodies in the proximal INL and ganglion cell layer were

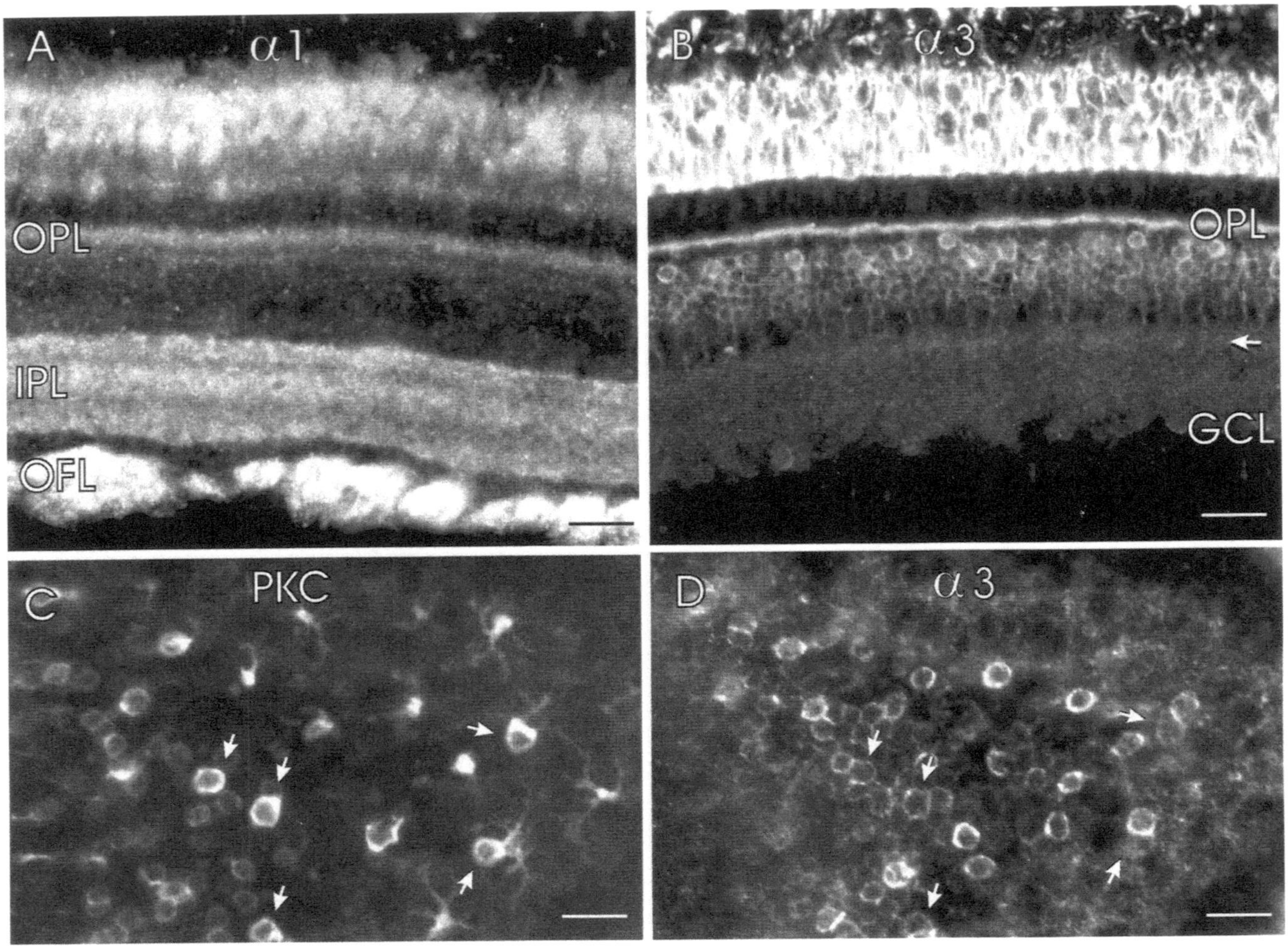

Fig. 8. $GABA_A$ receptor α-subunits. Scale bars = 20 μm. (A) $GABA_A\alpha_1$-IR was prominent in the OPL and IPL, in which it appeared as broad bands in layers 1, 3, and 5. There appeared to be a weak band of $GABA_A\alpha_1$-IR in the distal INL, perhaps associated with horizontal cells. The intense $GABA_A\alpha_1$-IR in the optic fiber layer (OFL) was likely to be non-specific because it decreased in intensity with decreases in primary concentration far more quickly than the label in the OPL and IPL. (B) $GABA_A\alpha_3$-IR was restrictedly to bipolar cell bodies and their processes in the OPL. Descending processes were only lightly labeled. There was a hint of label in layer 1 of the IPL (arrow) and in cells bodies in the ganglion cell layer. (C and D) Double-labeling with PKC (C) and $GABA_A\alpha_3$ (D) showed that only the large, bright PKC-IR bipolar cell bodies were weakly $GABA_A\alpha_3$-IR (arrows), but the smaller, dim PKC-IR cell bodies were not $GABA_A\alpha_3$-IR at all. The bright $GABA_A\alpha_3$-IR cells were not PKC-IR nor were any of the smaller $GABA_A\alpha_3$-IR cell bodies. Thus, $GABA_A\alpha_3$-IR was localized to the ON Mb bipolar cell bodies, OFF mixed bipolar cells and OFF cone bipolar cells. The ON cone bipolar cells were, by exclusion, not $GABA_A\alpha_3$-IR.

labeled lightly for VGAT. Although the IPL was labeled for VGAT throughout, there were three distinct laminas of puncta in layers 1, 3, and 5.

GABA receptors

We determined the localization of several subunits of the $GABA_A$ receptor and the ρ-subunit of the $GABA_C$ receptor (Fig. 8). The $GABA_A\alpha1$ subunit was most prominent in diffuse bands over the OPL, IPL and optic fiber layer (Fig. 8A). Labeling over the photoreceptors appeared to be mostly background. The band in the OPL was too diffuse to localize it to photoreceptor terminals or dendrites of second-order neurons. Just proximal to the OPL, there was a faint band that we have come to associate with the cell bodies of cone horizontal cells. No other cell bodies in the INL appeared to be labeled. The IPL was characterized by three broad bands in layers 1, 3, and 5, with clearly defined dark bands in layers 2 and 4. Layer 1 also had bright patches of $GABA_A\alpha1$ subunit labeling at the INL/IPL border. Although cell bodies in the GCL were not labeled at all, there was intense label over the optic fiber layer. With serial dilution, fluorescence over the photoreceptors and the optic fiber layer decreased more rapidly than over the OPL and IPL. The trilaminar pattern in the IPL was less obvious but the bright patches at the INL/IPL border remained.

The $GABA_A\alpha3$ subunit appeared to label every cell body in the mid INL and numerous cell bodies in the GCL (Fig. 8B). There was intense label in the OPL and a weak band in layer 1 of the IPL. The $GABA_A\alpha3$-IR in the INL included larger bright cell bodies and more numerous, smaller, more weakly labeled cells, reminiscent of PKC-IR. However, double-labeling with PKC showed that only the large, bright PKC-IR bipolar cell bodies were weakly $GABA_A\alpha3$-IR; however, the smaller, dim PKC-IR cell bodies were not $GABA_A\alpha3$-IR at all (Fig. 8C and D). The bright $GABA_A\alpha3$-IR cells were not PKC-IR nor were any of the smaller $GABA\alpha_3$-IR cell bodies. We know from views of vertical retinal sections of PKC-IR (Fig. 4A) that ON-bipolar cell bodies are proximal in the mid INL. The circular shape of the cell bodies eliminates horizontal cells from consideration. Thus, we conclude the $GABA_A\alpha3$-IR cell bodies in the more distal region of the INL belong to OFF bipolar cells. It seems likely then that all OFF bipolar cells (mixed rod/cone and cone types) were immunoreactive for the $GABA_A\alpha3$ subunit, whereas only the mixed ON M*b* type was weakly $GABA_A\alpha3$ -IR.

Two monoclonal antibodies against the $GABA_A\beta_2/\beta_3$ subunits: mAbs 62-3G1 and BD-17, were used. These antibodies reportedly recognize the same epitope and should produce the same labeling pattern (Ewert *et al.*, 1992). We found this to be the case in zebrafish retina (Fig. 9A and B). Both antibodies labeled only the IPL, with enhanced labeling at layers 1 and 3. Small puncta also could be seen scattered within layer 5. Labeling in the OPL was weak at best and did not appear to be much above background. This identity of label is different from reports in goldfish in which these antibodies were used (Yazulla *et al.*, 1989; Lin & Yazulla, 1994b). Both antibodies labeled the IPL with a similar pattern but only mAb 62-3G1 labeled the OPL. Another difference was that mAb 62-3G1 produced a considerable amount of somatic labeling in the INL, something not seen in zebrafish. Curiously, in cat and rabbit retinas, mAbs 62-3G1 and BD-17 do not produce the same pattern either (Hughes *et al.*, 1991; Brecha, 1992; Vardi *et al.*, 1992). Thus, zebrafish appears rather unique in this regard.

The $GABA_C\rho$-subunit was most prominent at the outer margin of the INL (Fig. 9C), with weaker punctate labeling throughout the IPL. It appeared as if the $GABA_C\rho$-subunit was surrounding horizontal cell bodies with processes extending into the OPL. This pattern is very different from that reported in goldfish (Koulen *et al.*, 1997) in which the $GABA_C$ ρ-subunit was restricted largely to the IPL, where it was particularly prominent surrounding the synaptic terminals of M*b* bipolar cells. Less intense but punctate labeling was seen in the outer retina. We cannot explain such differences between zebrafish and goldfish, except to note that the overall labeling pattern for the $GABA_C\rho$-subunit was similar for goldfish, chicken and cat (Koulen *et al.*, 1997) and quite different from that reported here in zebrafish.

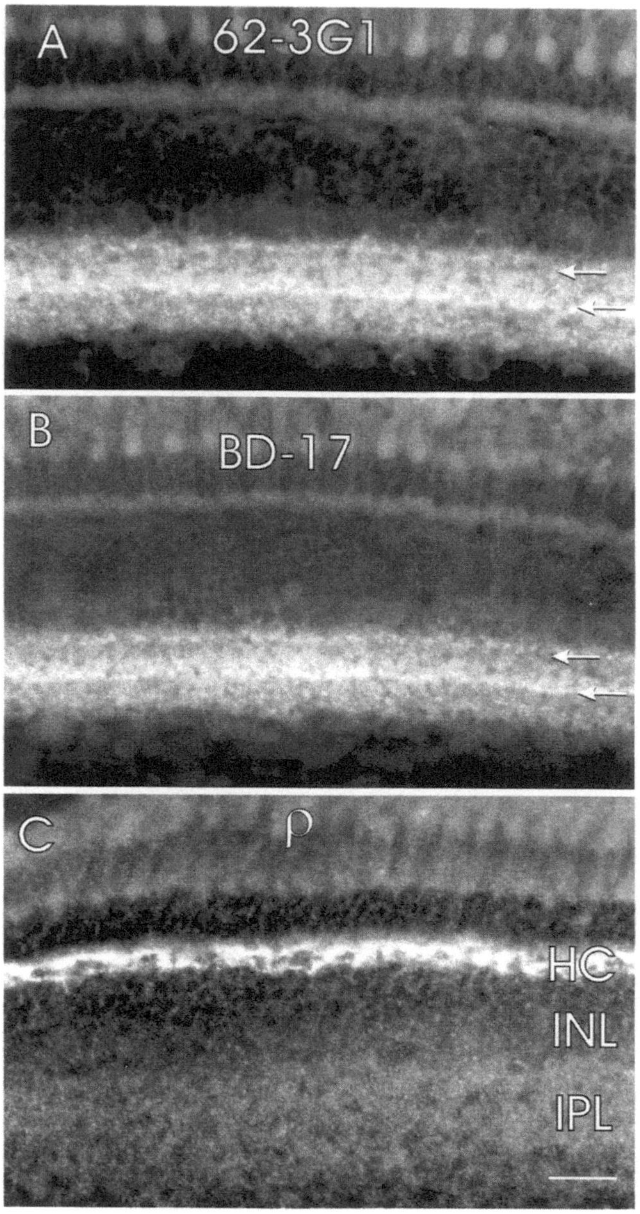

Fig. 9. $GABA_A$ receptors. Scale bars = 20 μm. (A) mAb 62-3G1 against the $GABA_A\beta_2/\beta_3$ subunits was restricted to the IPL, with brighter bands in layers 1 and 3 (arrows). Faint staining of amacrine cells and cells in the ganglion cell layer was observed as well. (B) mAb BD-17 against the $GABA_A\beta_2/\beta_3$ subunits had the identical distribution as mAb 62-3G1, throughout the IPL with bands in layers 1 and 3 (arrows). (C) The $GABA_C\rho$-subunit was most prominent in the outer margin of the INL, with weaker punctate labeling throughout the IPL. It appeared as if the $GABA_C\rho$-subunit was surrounding horizontal cell bodies (HC) with processes extending into the OPL.

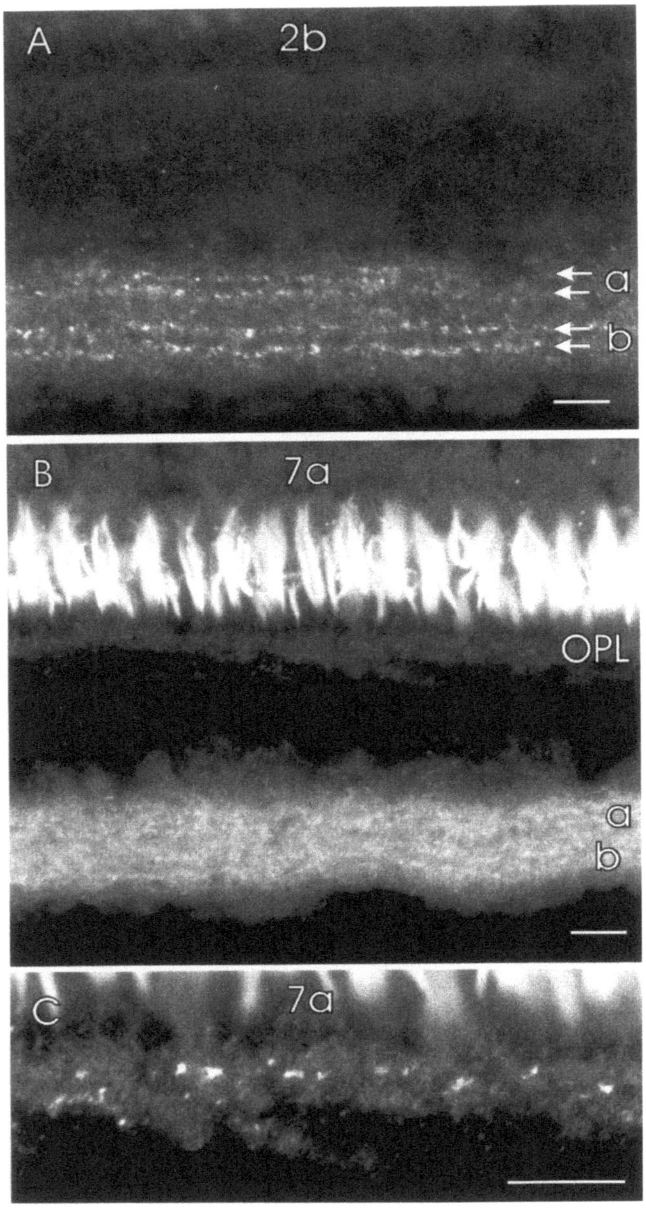

Fig. 10. Glycine receptors. Scale bars = 20 μm. (A) mAb 2B against the glycine receptor α-subunit was restricted to the IPL, appearing in four discrete laminas, two each in sublaminas *a* and *b*. (B and C) mAb 7a against gephyrin was concentrated in the IPL. Faint lamination was apparent in sublaminas *a* and *b*, similar to that seen in A. Faint label also was seen in the OPL, which in a higher magnified and enhanced image (C) appears as scattered boutons.

GLYCINERGIC SYSTEM

We used mAb 2B against residues 1–10 of the glycine receptor $\alpha 1$ subunit (Hoch *et al.*, 1991). Glycine receptor-IR (Fig. 10A) was restricted largely to the IPL, in which 4 laminas in layers 1, 2, 4, and 5 were embedded in a low level of overall label. Although puncta were seen in the vicinity of cell bodies in the proximal INL and ganglion cell layer, these did not appear to be cytoplasmic in origin. Occasionally, fine streaks of glycine receptor-IR were observed traversing the INL and the OPL. However, these were not nearly as prominent as the glycine receptor-IR in the IPL. This distribution is consistent with *in situ* hybridization studies that have cloned and localized zebrafish glycine receptor subunits $\alpha 1$ and β to the IPL (Imboden *et al.*, 2001); the $\alpha 2$ subunit was localized to both the OPL and IPL.

Gephyrin, was thought at first to be associated with chloride channels that were coupled to the strychnine-sensitive glycine receptor (Pfeiffer *et al.*, 1984; Altschuler *et al.*, 1986), but there has been growing evidence that gephyrin is associated with $GABA_A$-receptors too (*i.e.*, Triller *et al.*, 1987; Yazulla & Studholme, 1991a), particularly those containing α_2 and γ_2 subunits (*i.e.*, Sassoè-Pognetto *et al.*, 1995). The very intense gephyrin-IR of the cone axons is identical to that shown in goldfish and shown to be completely intracellular by EM-ICC (Yazulla & Studholme, 1991). In the more proximal retina, gephyrin-IR was concentrated in the IPL, with weakly-labeled boutons extending into the OPL, INL and ganglion cell layer (Fig. 10B). Although gephyrin-IR was very dense throughout the IPL, 4 laminas can be discerned in layers 1, 2, 4, and 5, identical to that observed with glycine receptor-IR. A higher magnification, contrast-enhanced view of the OPL shows gephyrin-IR boutons scattered in the OPL (Fig. 10C). This pattern is very similar to that observed in goldfish retina. The boutons in the OPL are likely to be on processes that are postsynaptic to Type II glycinergic interplexiform cells observed in goldfish (Marc & Lam, 1981; Yazulla & Studholme, 1991). However, in goldfish, lamination of gephyrin-IR was not observed in the IPL and punctate cytoplasmic label was observed more extensively in the perinuclear cytoplasm of cell bodies in the INL and GCL (Yazulla & Studholme, 1991).

GLUTAMATERGIC SYSTEM

Except for glutamate (Connaughton *et al.*, 1999; see also Marc & Cameron, 2001), kainate receptor subunits GluR5-7 (Nelson *et al.*, 2001) and glutamine synthetase (Peterson *et al.*, 2001) no other label for the glutamate transmitter system has been described in the zebrafish retina. Here we present data on a glutamate transporter GLT-1 and a variety of ionotropic and metabotropic glutamate receptors. Ionotropic glutamate receptors are of three kinds based on their pharmacological profiles, and each consists of multiple subunits. AMPA receptors consist of GluR1-4 subunits, kainate receptors consist of GluR5-7, KA1-2 subunits and NMDA receptors consist of NR1 and NR2 subunits.

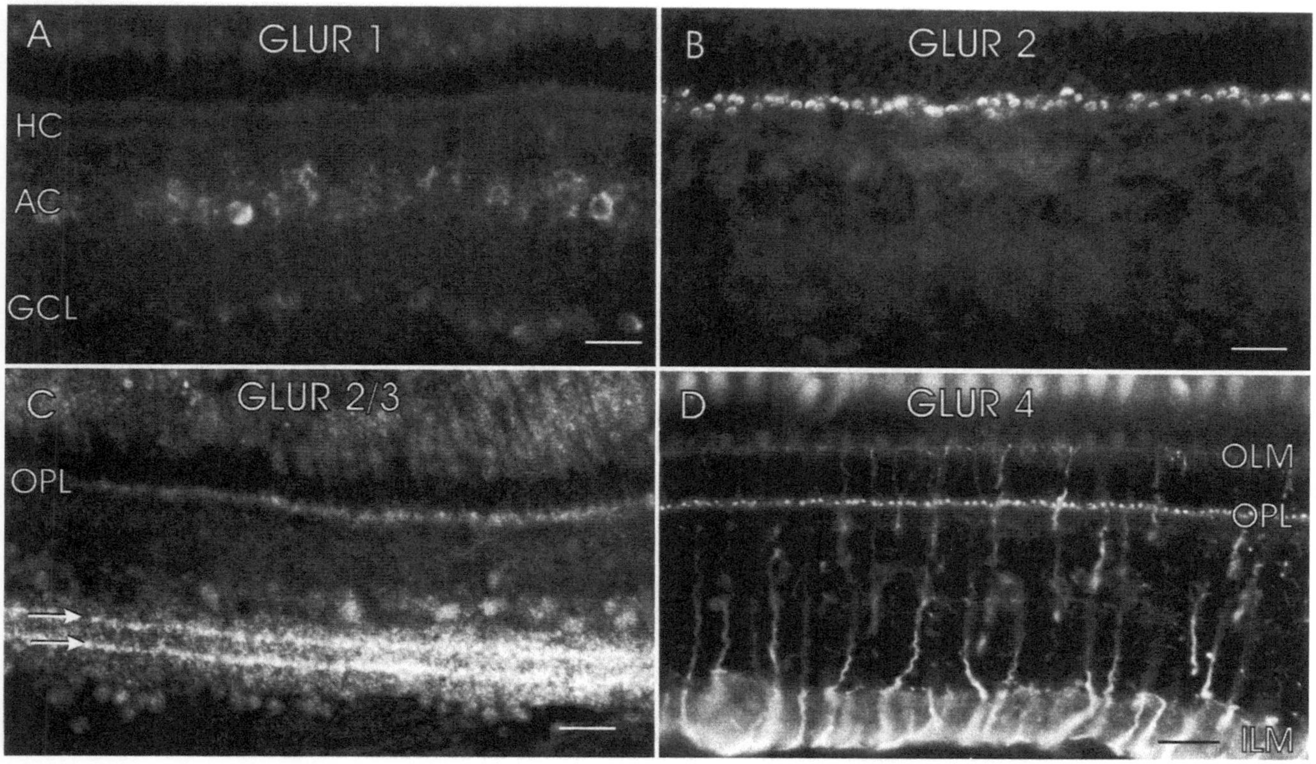

Fig. 11. AMPA glutamate receptors. Scale bars = 20 μm. (A) GluR1-IR was most prominent in amacrine cell bodies (AC) in the proximal INL and ganglion cell layer, with weak label over horizontal cell bodies (HC). (B) GluR2-IR was restricted to a thin layer in the OPL. The structures appeared as mushroom caps in a vertical view and as a rosette in an oblique view. (C) GluR2/3-IR was confined largely to the IPL, in which two prominent layers (2 and 4, arrows) were visible. Less dense labeling of cell bodies in the proximal INL, ganglion cell layer and processes in the OPL was observed. (D) GluR4-IR was prominent in Müller's cells, which were labeled completely from the inner to outer limiting membranes, including their cell bodies in the INL. There was also a band of fine puncta in the OPL that stood out brightly from the vertical streaks of Müller's cell processes in the area.

AMPA receptors

Each of the 4 AMPA receptor subunits produced a rather unique labeling pattern that together, resulted in relatively little labeling of the IPL (Fig. 11). GluR1 was unusual in that cell bodies rather than the synaptic layers were labeled. GluR1-IR was most prominent on cell bodies in the proximal INL and in the ganglion cell layer (Fig. 11A). Weak label was seen in the region of cone horizontal cell bodies in the distal INL. This pattern is similar to that reported in goldfish (VandenBranden *et al.*, 2000a) in that labeling was mostly somatic, although stronger than in zebrafish. The stock antiserum from Chemicon was used at 1:10 by us in zebrafish, but at 1:100 in goldfish. Also, our fixation was from 30 min to 60 min, a condition that reduced GluR1-IR in goldfish (VandenBranden *et al.*, 2000a). Recently, Schultz *et al.* (2001) used an antiserum against GluR1 from PharMingen at 1:500 in carp retina. They also report weak somatic labeling and no labeling in either plexiform layer. As there was no GluR1 detected in Western blots, Schultz *et al.* (2001) considered the GluR1-IR in the carp retina to be non-specific. Finally, GluR1 was not observed in goldfish by Peng *et al.* (1995), calling into question the validity of GluR1 in the fish retina.

GluR2 was unique in that label was confined to a narrow band in the OPL (Fig. 11B). GluR2-IR appeared as regularly spaced, dome-shaped structures in vertical section and as small rosettes when viewed in cross section. This is the same pattern reported in goldfish (VandenBranden *et al.*, 2000a; Klooster *et al.*, 2001). The GluR2-IR structures in goldfish were identified, by pre-embedding electron microscopic immunocytochemistry (EM-ICC), as belonging to the distal tips of rod and cone horizontal cells. All cones were innervated by GluR2-IR dendrites (Klooster *et al.*, 2001). Preliminary EM-ICC in this laboratory (Klooster & Yazulla, unpublished) confirms that GluR2-IR is localized to cone horizontal cell dendrites in the zebrafish as well. More will be said of this in later sections on double-labeling results.

Labeling by GluR2/3 was confined largely to the IPL, in which two prominent in layers (2 and 4) were visible (Fig. 11C). Less dense labeling of cell bodies in the proximal INL, ganglion cell layer and processes in the OPL was observed. This is very different from that reported in two studies in goldfish in which although cone terminals were most prominently labeled, there was lighter label of most of the remaining retina, particularly horizontal cells and the IPL (Peng *et al.*, 1995; VandenBranden *et al.*, 2000a). Schultz *et al.* (2001) used a GluR2/3 antiserum from PharMingen in carp retina and found a similar widespread distribution as in goldfish, including horizontal cells, some bipolar cells and their dendrites in the OPL, amacrine and ganglion cell bodies as well as numerous processes in the IPL. Since antisera against GluR2/3 cross react with epitopes on both GluR2 and GluR3, it should be possible to localize GluR3, by exclusion, to the bipolar cells, amacrine cells and the IPL, with the bulk of reactivity on horizontal cell dendrites due to GluR2. Of course, this remains to be determined directly by EM-ICC.

GluR4 produced the most extensive labeling of the AMPA receptor subunits (Fig. 11D). Müller's cells were labeled completely from the inner to outer limiting membranes, including their cell bodies in the INL. Also, present was a band of fine puncta in the OPL. These GluR4-IR puncta were smaller in dimension than those observed with GluR2-IR. This pattern is identical to that reported in goldfish and carp (Peng *et al.*, 1995; Schultz *et al.*, 1997; Yazulla & Studholme, 1999; VandenBranden *et al.*, 2000a; Schultz *et al.*, 2001). Processes in the OPL are likely due to the distal dendrites of OFF bipolar cells (Schultz *et al.*, 1997; Yazulla & Studholme, 1999; Schultz *et al.*, 2001) as determined by EM-ICC.

Kainate receptors

Contrary to the pattern observed for AMPA receptors, kainate receptors were most prominent in the IPL (Fig. 12). GluR5-7 had the most complicated pattern

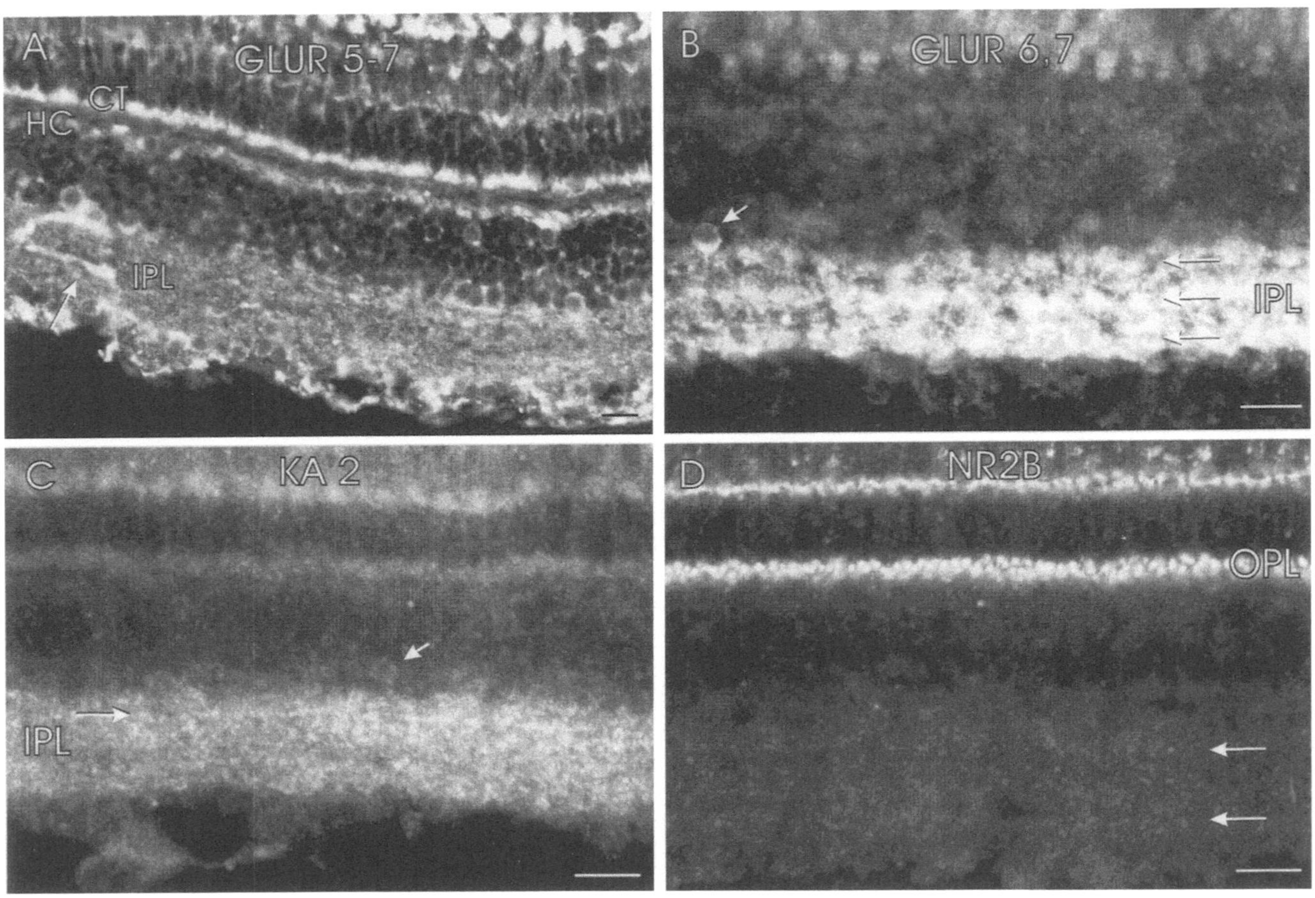

Fig. 12. Kainate and NMDA receptors. Scale bars = 20 μm. (A) GluR5-7 labeled cone photoreceptor terminals (CT), horizontal cell bodies (HC) and processes in the distal INL, large cell bodies in the proximal INL and large caliber processes in the distal half of the IPL (arrow). A general, more diffuse label was visible over the entire thickness of the IPL. (B) GluR6/7 was restricted to the IPL, in which it appeared as three broad laminas in layers 1, 3, and 5 (arrows), with sublamination in layer 5. Occasionally, a pyriform amacrine cell (short arrow) with a single descending process was labeled for anti-GluR6/7. (C) KA2-IR also was restricted to the IPL, but the pattern was finer than observed with GluR6/7. KA2-IR was observed in few amacrine cells (short arrow) and in a discrete band in layer 2 of the distal IPL (arrow). (D) NR2B-IR produced intense label over numerous, round structures in the OPL and weakly in two bands in layers 2 and 4 of the IPL (arrows).

with label over photoreceptor terminals, cell bodies and processes in the distal INL, large cell bodies in the proximal INL and large caliber processes in the distal half of the IPL (Fig. 12A). A general, more diffuse label was visible over the entire thickness of the IPL. It appeared as if horizontal cells as well as some bipolar cell bodies were labeled in the distal INL. The arborization of the large amacrine cells is curious; for in oblique sections, a large process entered the IPL, descended to the mid IPL and then branched in one direction in layer 3. The projection appeared to be polarized and restricted to the distal half of the IPL. The pattern of labeled cells in the outer zebrafish retina is consistent with a study on isolated zebrafish retinal neurons, in which GluR5-7-IR was found in all cones and horizontal cells; bipolar cells were labeled differentially with brighter label in large mixed type cells (Nelson *et al.*, 2001). This pattern is entirely different from that reported in goldfish and carp (VandenBranden *et al.*, 2000a; Schultz *et al.*, 2001) who used a different antibody: a mouse monoclonal from PharMingen whereas we used a mouse monoclonal from Chemicon. These authors reported relatively weak labeling of the OPL, but extensive labeling of horizontal cell axon terminals, punctate boutons in the IPL and vertical streaks in the IPL. We found that the Chemicon GluR5-7 antibody produced very weak labeling in goldfish, similar in pattern to zebrafish, when used identically as in zebrafish. Clearly, these two monoclonal antibodies are not recognizing the same antigen. This reinforces the need to properly characterize antibodies in the tissue under consideration when attempting to make functional statements based on immunocytochemistry. Without such characterization, we are not claiming to have presented the valid GluR5-7-IR pattern in zebrafish.

Labeling with anti-GluR6/7 was restricted to the IPL (Fig. 12B), in which it appeared as three broad laminas in layers 1, 3, and 5, with sublamination in layer 5. Occasionally, a pyriform amacrine cell with a single descending process was labeled for anti-GluR6/7. Superficially, this GluR6/7 pattern in the IPL resembles the GABA-T pattern, suggesting that GABAergic amacrine cells are enriched in the GluR6/7 subunits. This pattern is different than that reported in goldfish (Peng *et al.*, 1995) in which label was most prominent over the horizontal cell axon terminals. There was lighter label over horizontal cell bodies, the OPL and IPL.

KA2-IR also was restricted to the IPL (Fig. 12C), but the pattern was finer than observed with GluR6/7. KA2-IR was observed in few amacrine cells and in a discrete band in layer 2 of the distal IPL.

NMDA receptors

NR1 did not label the zebrafish retina at all, even though it produces intense label of the entire IPL in goldfish (VandenBranden *et al.*, 2000a). Identical, parallel processing of goldfish and zebrafish by us produced the same pattern in goldfish as previously reported, but no label at all in zebrafish, even if we increased the concentration of the NR1 antibody by 10-fold, or reduced the fixation time to 30 min. The NR2B antibody however, produced a beautifully restricted pattern in the zebrafish (Fig. 12D). Intense label was found over numerous, round structures in the OPL and weakly in two bands in layers 2 and 4 of the IPL. See double-label section for more details on the identity of the NR2B-IR structures.

Metabotropic receptors

The distributions of metabotropic subunits mGluR1α, mGluR2/3, and mGluR5 (Fig. 13) were very different from each other. There was a considerable amount of diffuse labeling of the nuclei of the photoreceptors, amacrine cells and ganglion cells with mGluR1α-IR (Fig. 13A). However, there was discrete punctate labeling of cell bodies in the mid INL, a portion of their descending axon, and dendrites in the OPL with mGluR1α-IR. The position of these cells was suggestive of the ON mixed rod/cone bipolar cells. Indeed, double labeling with PKC (Fig. 13A and C) showed this to be the case. All ON bipolar cells in the zebrafish retina cone type and mixed rod/cone type are labeled by PKC-IR (Fig. 4A). Note, however, that mGluR1α-IR was restricted to the large bright PKC-IR mixed type. This is the identical pattern, including the co-localization with PKC-IR in only the ON mixed bipolar cells that was observed in goldfish (Yazulla *et al.*, 2001). Klooster *et al.* (2001) also showed by EM-ICC that mGluR1α-IR was localized to the dendrites of ON bipolar cells, innervating both rods and cones. It is highly likely that the same would be found in zebrafish.

mGluR2/3-IR was broadly distributed in the retina (Fig. 13B), superficially resembling that obtained with GluR5-7. mGluR2/3-IR was prominent over photoreceptor terminals, elongated cell bodies in the distal INL, few amacrine cells in the proximal INL, two bright laminas in layers 1 and 5 of the IPL and a lower density in the middle of the IPL. The labeled cell bodies in the distal INL appear to be a second layer of horizontal cell bodies because the empty region that is distal to them is likely to be occupied with the GAD-IR horizontal cell bodies (see Fig. 5). The differential labeling of horizontal cells by GAD-IR and mGluR2/3-IR remains to be determined.

The distribution of mGluR5-IR was the most restricted of the subunits (Fig. 13D), being confined to the synaptic terminal of cones. The label appeared as very fine grains over the cone pedicles. There did not appear to be any labeling over the region of rod spherules, just distal to the cone pedicles. It also was not clear if all cone pedicles were mGluR5-IR.

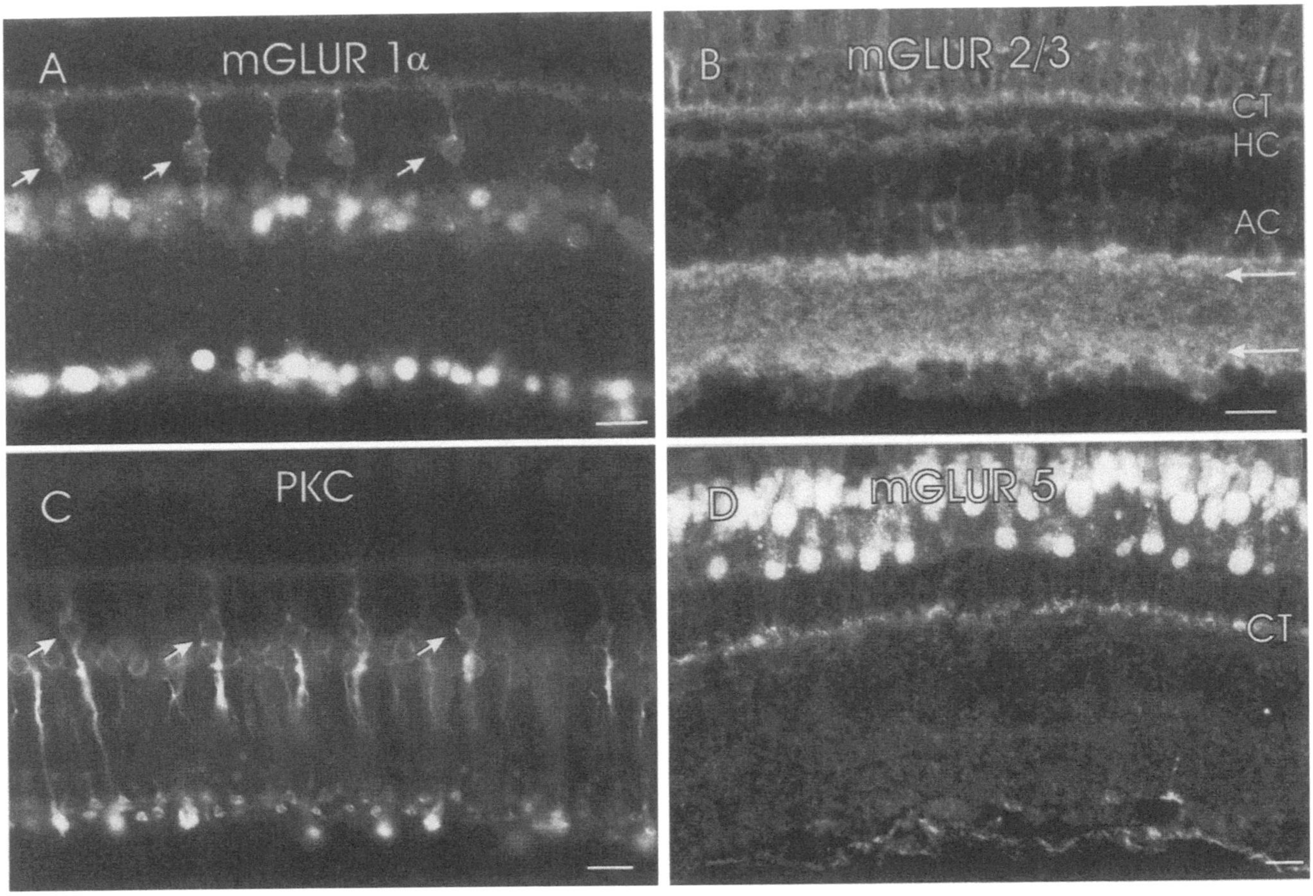

Fig. 13. Metabotropic glutamate receptors. Scale bars = 20 μm. (A and C) mGluR1α—There was a considerable amount of background labeling of the nuclei of the amacrine cells and ganglion cells. However, there was discrete punctate labeling of cell bodies in the distal half of the INL, a portion of their descending axon, and dendrites that extended laterally in the OPL. Double labeling with PKC-IR (C) showed that only the large PKC-IR cells co-localized with mGluR1α (arrows), showing that mGluR1α was restricted to the large Mb bipolar cells. (B) Labeling with mGluR2/3-IR was prominent over cone photoreceptor terminals (CT), horizontal cell bodies (HC), few amacrine cells in the proximal INL (AC), two bright laminas in layers 1 and 5 of the IPL (arrows) and a lower density in the middle of the IPL. The labeled cell bodies in the distal INL appear to be a second layer of horizontal cell bodies. (D) The distribution of mGluR5-IR was the most restricted of the subunits, being confined to the synaptic terminal of cones (CT). The label appeared as very fine grains over the cone pedicles. It was not clear if all cone pedicles were mGluR5-IR.

Glutamate transporter

GLT-1-IR was very intense in a narrow band in the OPL (Fig. 14), with much lighter label in the IPL. Despite the light label in the IPL, there was a distinct dark band at about layer 3, the same region observed for the dark band with the GAD antisera (see Fig. 5). The intense puncta in the OPL were in the region of cone pedicles. However, despite the light GLT-1-IR over the cone axons, inner and outer segments, it is not clear if the GLT-1-IR in the OPL was in the cone pedicle or processes invaginating the pedicle. This pattern in zebrafish is quite different from that reported in goldfish (VandenBranden *et al.*, 2000b), in which the most prominent label was on Müller's cells with lighter label in Mb bipolar cells, the OPL and IPL. EM analysis showed membrane-bound GLT-1-IR at the base of cone pedicles and bipolar cell terminals in goldfish. It is likely that the GLT-1-IR in the zebrafish OPL is associated with cone terminals (see Double Label section), although that identity is not definite. There was also labeling along the basal lamina at the vitreal margin. Similar labeling in goldfish retina (VandenBranden *et al.*, 2000) persisted following preadsorption of the primary antibody with the peptide antigen, indicating that this labeling was non-specific.

BIOGENIC AMINES

In this section we report on the distribution of enzymes for catecholamine biosynthesis: tyrosine hydroxylase (TOH), dopamine beta hydroxylase (DBH) and phenylethanolamine-N- methyltransferase (PNMT), dopamine D1 receptor, serotonin

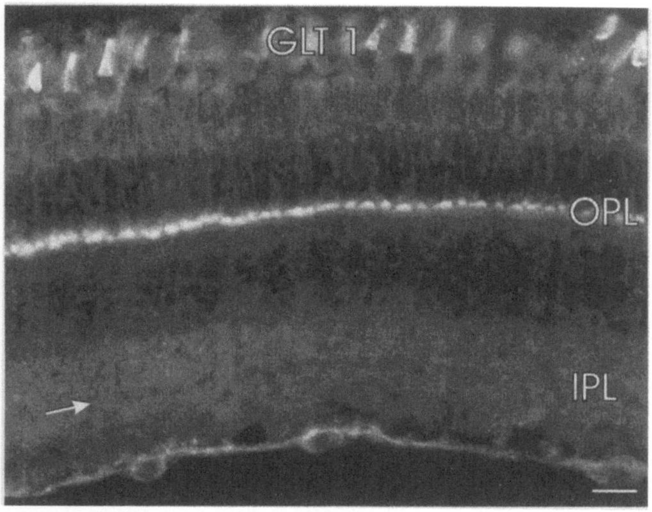

Fig. 14. Glutamate transporter—GLT-1-IR was very intense in a narrow band in the OPL, extending along the cone axons to the cone cell bodies. The intense puncta in the OPL are in the region of cone pedicles. Despite the light label in the IPL, there was a distinct dark band at about layer 3 (arrow), the same region observed for the dark band with the GAD antisera (see Fig. 5). Scale bar = 20 μm.

(5 hydroxytryptamine—5HT) and the catabolic enzyme for adenosine, adenosine deaminase (ADA).

Catecholamine enzymes

The distribution of TOH-IR in zebrafish retina (Fig. 15A) is very similar to that observed in goldfish (Yazulla & Zucker, 1988; Van Haesendonck & Marc, 1993). These appeared to be typical interplexiform cells (IPC) of fish retina. There are very few, large cell bodies in the proximal INL that give rise to processes that ramify in the IPL and OPL. Although TOH-IR processes were present throughout the IPL, they were more numerous in the proximal portion. Processes in the OPL were punctate and surrounded elongated cell bodies in the distal INL, likely to be horizontal cells. Compared with goldfish that had a very clean pattern only over the IPCs, zebrafish retina had a higher overall background with labeling of blood vessels at the vitreal margin and cone axons. We were surprised that the mouse mAb αTH against quail tyrosine hydroxylase from the Developmental Studies Hybridoma Bank did not label either the zebrafish or goldfish retina because this antibody produced beautifully clean labeling of the large dopamine amacrine cells in the salamander retina (Yang, personal communication).

DBH-IR was not found in the zebrafish retina, consistent with our own unpublished screening in goldfish and the work of Foster *et al.* (1985) in rat retina. DBH-IR was present in numerous fibers in the choroid, ciliary body and iris (not shown). PNMT-IR, however, was distributed widely in the zebrafish retina (Fig. 15B). It was most intense in three broad bands at layers 1, 3 and 5 in the IPL, cell bodies in the proximal INL and in a narrow band in the OPL. It is possible that the PNMT-IR in the OPL was due to processes of Type I3-IPCs as described in goldfish retina (Baldridge & Ball, 1993), rather than horizontal cells as described for the ferret and rat retinas (Keyser *et al.*, 1987; Hammang *et al.*, 1992). PNMT-IR

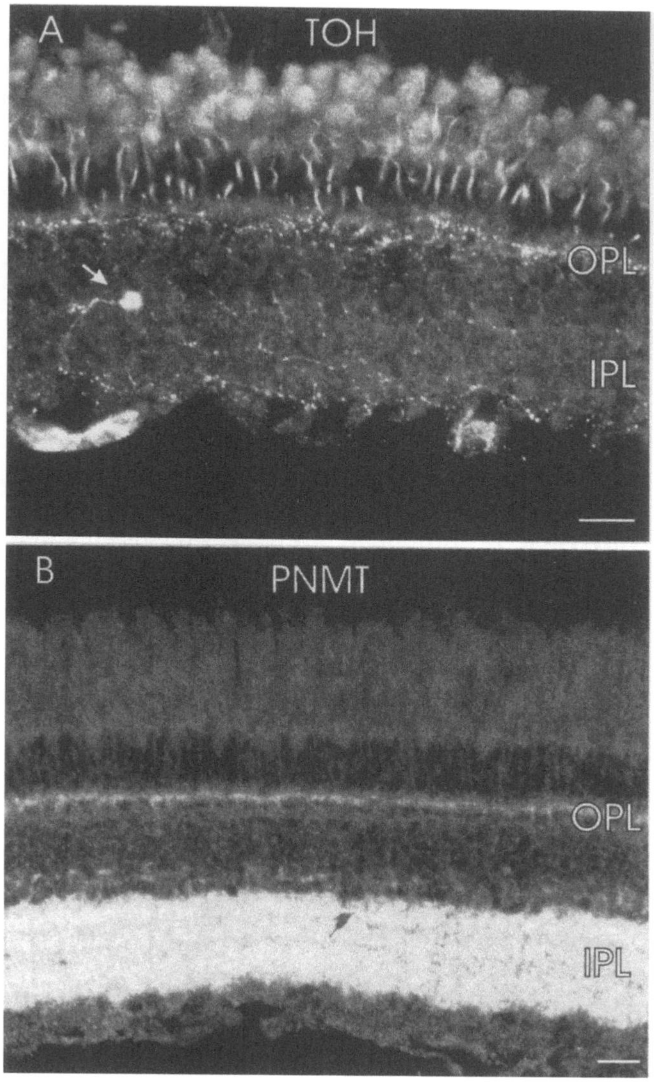

Fig. 15. Catecholamines. Scale bars = 20 μm. (A) Tyrosine hydroxylase—There are very few, large cell bodies in the proximal INL (arrow) that gave rise to processes that ramified in the IPL and OPL. Although TOH-IR processes were present throughout the IPL, they were more numerous in sublamina b. Processes in the OPL were punctate and surrounded elongated cell bodies in the distal INL, likely to be horizontal cells. Labeling over the photoreceptors and vitreal margin was not specific. (B) PNMT-IR was distributed widely but was most intense in three broad bands at layers 1, 3 and 4/5 in the IPL. Amacrine cell bodies in the proximal INL appeared more intensely labeled than cells in the mid INL. The narrow band of label in the OPL is of an indeterminate origin.

in the zebrafish IPL was very intense, with 3 broad laminas in layers 1, 3 and 4/5, far more extensive than reported in goldfish retina (Baldridge & Ball, 1993).

The curious nature of these patterns is in the widely accepted sequence for catecholamine biosynthesis: TOH for dopamine, DBH for norepinephrine and PNMT for epinephrine. PNMT should not be distributed more widely than TOH and where is the DBH that would provide norepinephrine as the precursor for epinephrine synthesis by PNMT? These differences have been noted in the mammalian central nervous system (Foster *et al.*, 1985; Nguyen-Legros *et al.*, 1986; Keyser *et al.*, 1987; Hammang *et al.*, 1992) and it is likely that it is not due to cross-reactivity of antisera. We are not attempting to solve the problem here, but merely point out the necessity to search for an alternative pathway for epinephrine synthesis that bypasses TOH and DBH. A recent addition to this issue is the finding that TOH-IR and the dopamine transporter (*dat*) co-localize completely in the retina of larval zebrafish (Holzschuh *et al.*, 2001). The frequency and distribution of the TOR-IR/*dat* neurons was far less than that which we observed with PNMT-IR, again suggesting an alternative pathway for epinephrine synthesis.

Dopamine D1 receptor

The D1R antiserum used in this study was generated against a 18 amino acid peptide from the C-terminus of the goldfish D1R that shares a 50% homology with the rat and human D1R (Mora-Ferrer *et al.*, 1999). The goldfish D1R is truncated in the C-terminus and thus, D1R antisera against a mammalian D1R C-terminus do not crossreact with goldfish or zebrafish. D1R-IR was not particularly intense, but was enriched in the OPL, cell bodies in the proximal INL and throughout the IPL (Fig. 16A). Of particular note were patches of D1R-IR spaced regularly in the proximal IPL. Labeling of tissue at the vitreal margin is observed with numerous other antisera, and it is difficult to ascribe any importance to it. However, the remaining label as well as label over the cone axons is similar to that observed in the goldfish retina with this antiserum (Mora-Ferrer *et al.*, 1999). In goldfish retina, D1R-IR co-localizes with PKC on the cell body and synaptic terminal of mixed M*b* rod/cone bipolar cells. In zebrafish however, we could detect co-localization only over the synaptic terminal of the M*b* bipolar cells (Fig. 16A and B). The distribution of D1R-IR in the IPL is far more extensive than the distribution of dopamine release sites as indicated by TOH-IR, suggestive of a paracrine action of dopamine as discussed for goldfish (Yazulla & Studholme, 1995, 1997).

Serotonin (5HT)

First, we note that there was no preloading of the retina with serotonin or 5,7-dihydroxytryptamine, which is required to detect 5HT-IR in many other preparations, but not goldfish (Marc *et al.*, 1988). The distribution of 5HT-IR (Fig. 17A) was very similar to that previously described for goldfish (Marc *et al.*, 1988). There appeared to be a single type of amacrine cell (S1) that had bulbous terminals that ramified in layers 1 and 5 of the IPL with more scattered processes in the mid IPL. The cell bodies could appear at various places at the proximal margin of the INL, sometimes protruding into the IPL or displaced slightly into the INL. However, the axonal arborization gave no indication that these represented multiple subtypes. Marc *et al.* (1988) showed that multiple cell types took up ^{3}H-serotonin, but only the S1 type was 5HT-immunoreactive without preloading. Teranishi *et al.* (1987) reported two subtypes of 5HT-IR amacrine cells in the carp retina. However, the second type was far more prominent in retinas that were preloaded with 5,7-dihydroxytryptamine.

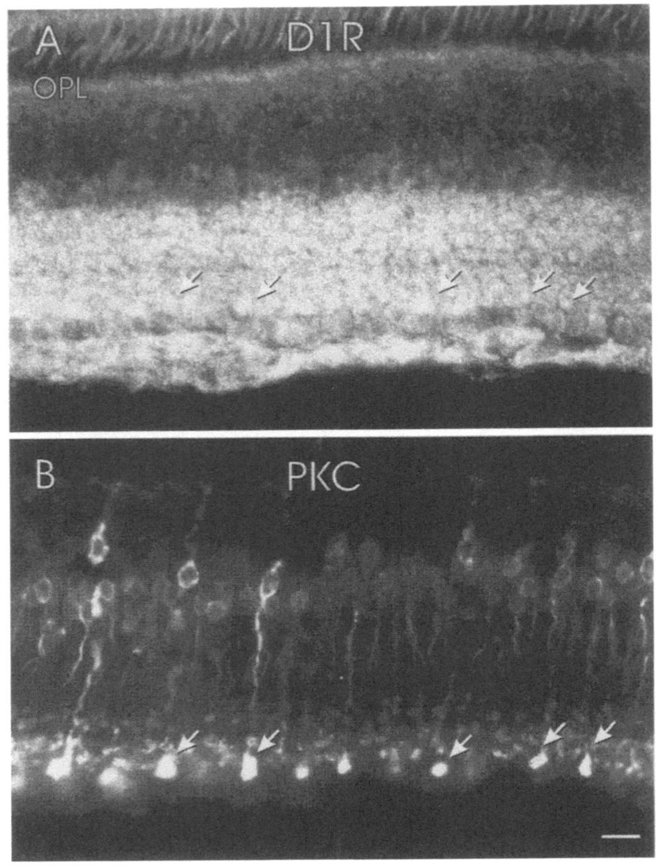

Fig. 16. Dopamine receptor—D1R-IR was weak in the OPL, cell bodies in the proximal INL and prominent throughout the IPL. Of particular note were patches of D1R-IR spaced regularly in the proximal IPL, that co-localized with PKC-IR (A, B arrows), demonstrating the D1R-IR was on the synaptic terminal of mixed Mb rod/cone bipolar cells. Scale bar = 20 μm.

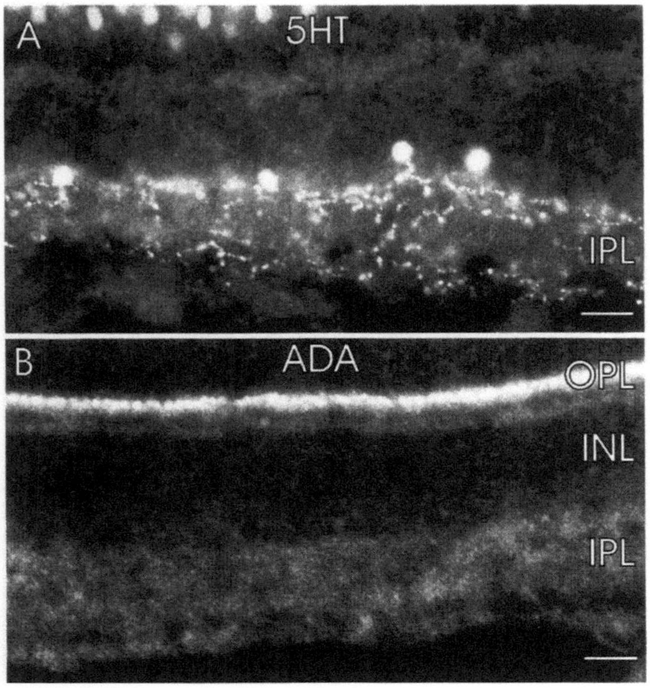

Fig. 17. Biogenic amines. Scale bars = 20 μm. (A) Serotonin—There appeared to be a single type of 5HT-IR amacrine cell that had bulbous terminals that ramified in layers 1 and 5 of the IPL with more scattered processes in the mid IPL. The cell bodies appeared at various places at the proximal margin of the INL, sometimes protruding into the IPL or displaced slightly into the INL. However, the axonal arborization gave no indication that these represented multiple subtypes. (B) Adenosine deaminase—ADA-IR was most prominent in a very bright band in the OPL. Much less intense label was observed in the IPL, amacrine cell bodies in the proximal INL and in a narrow band in the distal INL. This latter band appeared to be in the region of cone horizontal cell bodies.

Adenosine deaminase

Adenosine deaminase (ADA) is the catabolic enzyme that converts adenosine into inosine when high levels of adenosine saturate the kinases that phosphorylate adenosine into nucleotides. There is evidence that both adenosine and ATP are transmitters in the retina (*i.e.*, Blazynski & Perez, 1992; Neal & Cunningham, 1994; Neal *et al.*, 1998; Santos *et al.*, 2000). Once released, ATP would be converted into adenosine by ectonucleotidases. Indeed, high levels of ^{3}H-ATP-derived uptake and ^{3}H-adenosine-derived uptake have been shown in fish horizontal cells (Ehinger & Perez, 1984; Young & Dowling, 1989; Studholme & Yazulla, 1997). In zebrafish, ADA-IR was most prominent in a very bright band in the OPL (Fig. 17B). Much less intense label was observed in the IPL, amacrine cell bodies in the proximal INL and in a narrow band in the distal INL. This latter band appeared to be in the region of cone horizontal cell bodies. This pattern is different than that described for goldfish (Studholme & Yazulla, 1997), in which ADA-IR is most prominent on horizontal cell bodies, particularly the rod horizontal cells that showed the most intense ^{3}H-adenosine derived uptake. The intensity of ADA-IR in the goldfish OPL was weak at best, compared to the cell bodies. This is the reverse of what was observed in zebrafish. The section on Double-Labeling provides additional information on the identity of the ADA-IR structures in the zebrafish OPL.

CHOLINERGIC SYSTEM

Choline acetyltransferase

We tried two antisera against ChAT. The goat-polyclonal and mouse monoclonal antibodies produced the same pattern in the inner retina (Fig. 18), which was quite different from that reported in any other vertebrate. ChAT-IR was found intensely in numerous amacrine cell bodies in the proximal INL but only weakly in cell bodies in the ganglion cell layer. There were four laminas in the IPL, layers 1, 2 and 4, 5. Layers 2 and 4 were denser and appeared somewhat continuous, whereas layers 1 and 5 were lighter and appeared as broken clusters. Amacrine cells in the INL appeared to project to all four layers, as illustrated for a pyriform amacrine cell in Figure 18B. This is in marked contrast to the pattern in mammals in which amacrine cells and displaced amacrine cells are made up of mirror-image populations that project to a distal layer 2 and proximal layer 4, respectively. The presence of ChAT-IR in layers 1 and 5 has not been reported

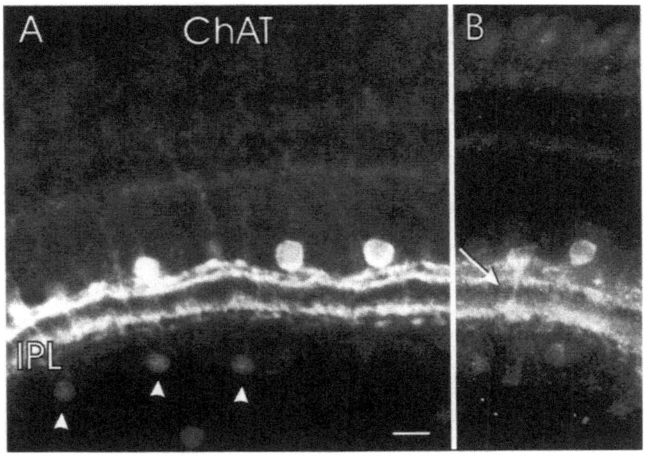

Fig. 18. Choline acetyltransferase—ChAT-IR was found in numerous amacrine cell bodies in the proximal INL but only weakly in cell bodies in the ganglion cell layer (arrowheads). There were four laminas in the IPL, layers 1, 2 and 4, 5. Layers 2 and 4 were denser and appeared somewhat continuous, whereas layers 1 and 5 were lighter and appeared as broken clusters. Amacrine cells in the INL projected to all four layers, as illustrated for a pyriform amacrine cell (arrow in B). Scale bar = 20 μm.

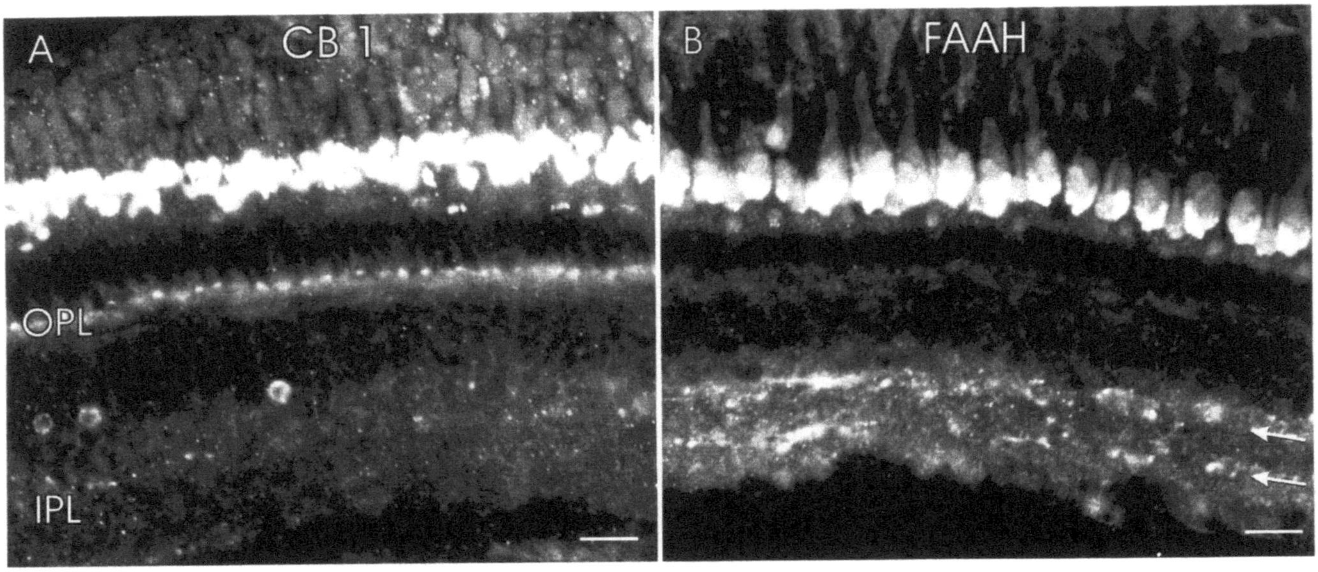

Fig. 19. Cannabinoids. Scale bars = 20 μm. (A) Cannabinoid CB1-receptor—CB1-IR was characterized by discrete puncta in the OPL, scattered amacrine cell bodies in the proximal INL, and sparse processes in a thin lamina in the distal IPL. The intense label in the ONL was not specific. (B) Fatty acid amide hydrolase—FAAH-IR was most prominent in two laminas at about layers 1 and 4 of the IPL (arrows), although there was lighter label throughout the IPL. The FAAH-IR fibers were larger in diameter than the CB1-IR fibers.

before in goldfish (Tumosa *et al.*, 1984; Tumosa & Stell, 1986) and at first we thought this was a unique difference between goldfish and zebrafish. However, Marc (personal communication) told us he observed a similar pattern in goldfish and we then corroborated his observation with the goat polyclonal anti-ChAT. Thus, both goldfish and zebrafish show four layers of ChAT-IR in the IPL and sets them apart from other species that show mirror-image cholinergic amacrine cell populations.

The monoclonal antibody mAb270 against the neuronal nictotinic acetylcholine receptor produced no labeling in the zebrafish retina.

CANNABINOIDS

Recently, cannabinoid CB1 receptors have been described in several vertebrate retinas, including goldfish (Straiker *et al.*, 1999; Yazulla *et al.*, 1999, 2000). CB1-IR, in the zebrafish retina (Fig. 19A) was characterized by discrete puncta in the OPL, scattered amacrine cell bodies in the proximal INL, and sparse processes in a thin lamina in the distal IPL. Again, the label in the cone inner segments is typical of many antisera and we do not present that label as specific. This pattern in the OPL is similar to goldfish in that the puncta in the OPL may be due to patches of CB1-IR at the base of cone pedicles (Yazulla *et al.*, 2000). The faint glow of CB1-IR of cone axons supports the photoreceptor location of CB1-IR, but ultrastructural analysis is required to confirm this. This pattern differs from goldfish in that there was no obvious label over Müller's cells or on the synaptic terminal of bipolar cells.

The enzyme for endocannabinoid degradation (*i.e.*, arachidonylethanolamide [anandamide], and 2-arachidonylglycerol [2-AG]) is fatty acid amide hydrolase (FAAH) (Di Marzo & Deutsch, 1998; Mechoulam *et al.*, 1998, for reviews). FAAH-IR appears to be restricted to two laminas at about layers 1 and 4 of the IPL (Fig. 19B). The FAAH-IR fibers were larger in diameter than the CB1-IR fibers. The only other reports of FAAH-IR in the retina are in rat (Yazulla *et al.*, 1999) and goldfish (Studholme *et al.*, 2000). In both species, particularly in the rat, the distributions of CB1-IR and FAAH-IR appear more complimentary than overlapping. In goldfish, CB1-IR and FAAH-IR are co-localized in cone photoreceptors, but they do not co-localize in bipolar cells. There appears to be little relationship between the location of CB1-IR and FAAH-IR in the zebrafish. Again, we do not make much of the FAAH-IR in the cone inner segments. Perhaps other receptors, for example the vanilloid receptors, will be more broadly distributed in the IPL, and thus, more closely resemble FAAH-IR.

NEUROPEPTIDES

We surveyed a representative sample of neuroactive peptides that have been localized in a wide variety of retinas. Curiously, neither glucagon- nor neurotensin-immunoreactivies were found in the zebrafish retina even though these have been reported in goldfish (Tornqvist & Ehinger, 1983; Li *et al.*, 1986) by immunocytochemistry, and glucagon by RIA (Ekman & Tornqvist, 1985). See Marshak (1992) for an excellent review of the anatomy and physiology of neuropeptides in fish retinas.

Although VIP increases adenylate cyclase activity in isolated horizontal cells of carp retina (Lasater *et al.*, 1983; Watling & Dowling, 1983), VIP-IR has not been observed in goldfish or elasmobranch retinas (Brecha, 1983; Bruun *et al.*, 1985), although a very low level of VIP was detected by RIA in goldfish (Ekman & Tornqvist, 1985). Recently, Mathieu *et al.* (2001) reported the presence of VIP by immunocytochemistry and immunoblot in the developing central nervous system, including the retina, of zebrafish. These authors used a rabbit anti-porcine VIP from Biomeda, CA. Data for the adult zebrafish retina were not provided and it would be of interest to determine if VIP-IR could be detected in other fish retinas as might be expected from the positive effects of VIP on adenylate cyclase activities in fish retinas.

Cholecystokinin (CCK)

CCK-IR was found at low frequency in regularly spaced amacrine cell bodies at the proximal margin of the INL (Fig. 20A). Their processes formed two diffuse bands at about layers 2 and 4 of the IPL. Despite relatively numerous reports of CCK-IR in the retinas of other species (*i.e.*, Osborne, 1981, 1982), this is the first report of which we are aware, on the distribution of CCK-IR in a fish retina. In all cases, CCK-IR is found in relatively rare populations of amacrine cells with a sparse distribution in the IPL.

FMRFamide (FMRF)

FMRF-IR was most prominent in a single layer of bright boutons that were confined to layer 1 of the IPL, which in horizontal sections appeared as a loose network of processes (Fig. 20B and C). No labeled cell bodies were ever observed anywhere in the retina. On very rare occasions, processes were observed to extend from the ganglion cell layer to the main plexus in the IPL, and from the IPL towards the OPL. These processes were likely derived from retinal efferents originating in the olfactory bulb and hypothalamus as reported in zebrafish (Pinelli *et al.*, 2000) and as first described in goldfish retina (Stell *et al.*, 1984; Kawamata *et al.*, 1990).

Enkephalin

Leu-Enk and Met-Enk were among the more difficult antisera to use in the zebrafish retina. The signal in the IPL, in particular, was very weak for both antibodies. First, for Leu-Enk, three types of amacrine cell were identified (Fig. 21A and B): a fusiform cell in the INL with processes in layers 1 and 5, a pyriform cell in the INL with processes in layer 5 and a cell in the ganglion cell layer with processes in layer 5. For Met-Enk, a single type of amacrine cell with processes in layers 1

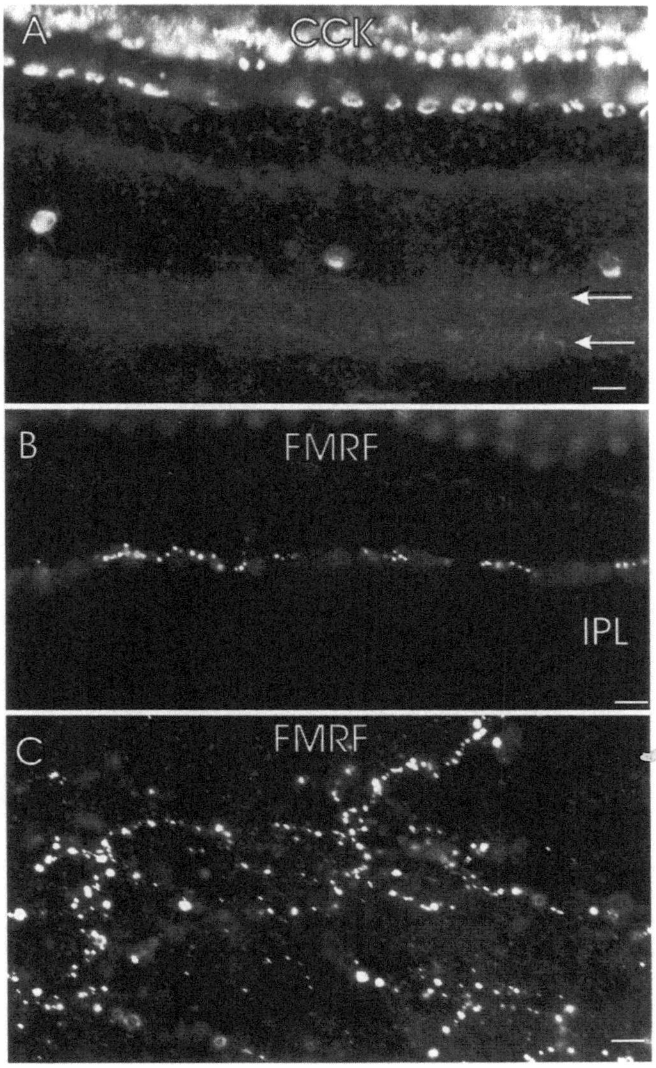

Fig. 20. Neuropeptides. Scale bars = 20 μm. (A) Cholecystokinin—CCK-IR was found at low frequency in regularly spaced amacrine cell bodies at the proximal margin of the INL. Their processes formed two diffuse bands at about layers 2 and 4 of the IPL (arrows). Labeling over the photoreceptors was not specific. (B and C) FMRFamide—FMRF-IR was most prominent in a single layer of bright boutons that were confined to layer 1 of the IPL, which in horizontal sections appeared as a loose network of processes (C). No labeled cell bodies were ever observed anywhere in the retina.

and 5 was labeled (Fig. 21C). The processes in the IPL were sparse for both Leu-Enk and Met-Enk, although more difficult to observe with Met-Enk. Higher concentrations of both antisera resulted in labeling of cone terminals. This was accompanied by intense labeling of the cone inner segments for Leu-Enk and thus its significance is uncertain. However, for Met-Enk, the labeling of the cone photoreceptors was restricted to the pedicles and, as such, may be more meaningful. Higher magnification (Fig. 21D) shows the Met-Enk-IR to be granular and confined to the pedicles. The patterns shown

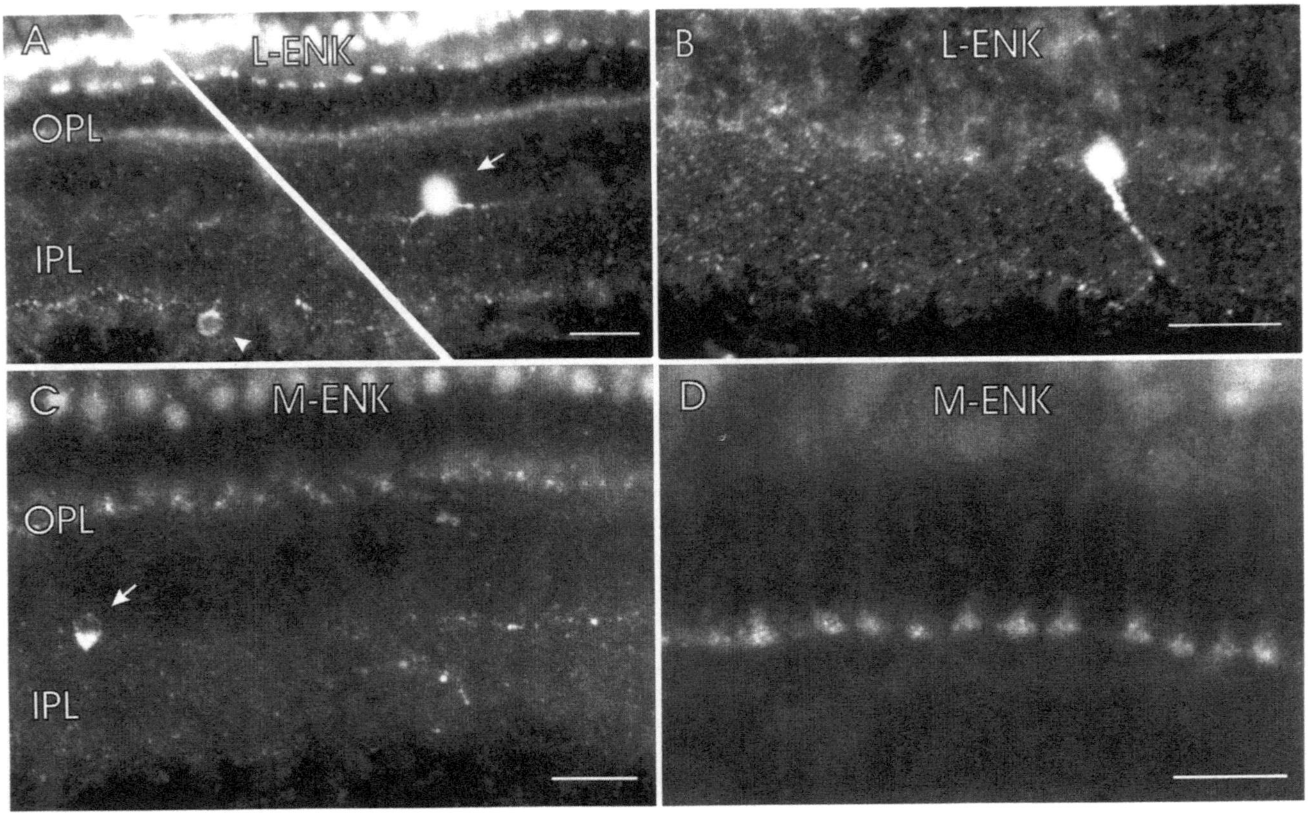

Fig. 21. Enkephalins. Scale bars = 20 μm. (A and B) Leu-Enk—three types of amacrine cell were identified: a fusiform cell in the INL with processes in layers 1 and 5 (A, arrow), a pyriform cell in the INL with processes in layer 5 (B) and a cell in the ganglion cell layer with processes in layer 5 (A, arrowhead). Diffuse labeling in the OPL was equivocal. "A" is a montage of the same section taken at different focal planes. (C and D) Met-Enk,—a single type of amacrine cell with processes in layers 1 and 5 was labeled (C, arrow). The labeling of the cone photoreceptors was restricted to the pedicles and, at higher magnification shows the Met-Enk-IR to be granular and confined to the pedicles (D).

for zebrafish are comparable to that shown in goldfish retina by Su *et al.* (1986) who used antisera against Leu-Enk that also cross-reacted with Met-Enk.

Neuropeptide Y (NPY)

NPY-IR was characterized by bistratified amacrine cells with cell bodies at the proximal margin of the INL that gave rise to two dense plexes in layers 1 and 5 of the IPL (Fig. 22A). Occasional boutons were observed in the mid IPL. The process connecting the laminas in the IPL appeared to be thinner than the processes comprising either of the arborizations. This pattern is very similar to that reported in a wide variety of teleost fishes: goldfish, blue acara, killifish or rainbow trout (Osborne *et al.*, 1985; Muske *et al.*, 1987; Negishi & Wagner, 1995; Subhedar *et al.*, 1996; Blank *et al.*, 1997). The main difference is the relative densities of the two plexes: the trout appears less intense than the zebrafish, whereas the killifish appears to have more intensely labeled processes. This, of course, could be due just to differences in tissue preparation and antisera concentrations.

Substance P

Substance P-IR was found in a single type of pyriform amacrine cell that sent a single process into the IPL and ramified exclusively in lamina 3 (Fig. 22B). The cell bodies were rare; usually only two or three were visible in a 3 mm section of the retina. This appearance is the same as that described years ago in the goldfish retina (Brecha *et al.*, 1981; Yazulla *et al.*, 1985). Li *et al.* (1986) referred to this cell as an SP-1 type and showed that it contained both substance P-IR and neurotensin-IR. The SP-1 cell was differentiated from an SP-2 type that had processes restricted to layer 1 of the IPL. The SP-2 cell did not stain for neurotensin-IR, but it was only observed using an uncharacterized polyclonal antiserum. As mentioned before, zebrafish retina did not stain for neurotensin. And it is thus possible that the zebrafish retina does not contain a counterpart to the SP-2 cell found in goldfish.

Somatostatin (SRIF)

We identified three types of amacrine cell that were SRIF-IR (Fig. 22C and D). These appear identical to those reported in goldfish (Yamada *et al.*, 1980) and their nomenclature will be used here. The SA*a*1 and SA*a*2 types each had their processes restricted to layer 1 of the IPL. However, the SA*a*1 type had larger cell body and smooth dendrites whereas the SA*a*2 type had a small cell body and varicose processes. The Sa*b* type was a pyriform cell that gave rise to a thin process that ramified in layer 5 of the IPL. A comparable cell in goldfish has a similar appearance to the pyriform A*b* GABA-accumulating amacrine cell; however, these cells come from separate populations (Yazulla *et al.*, 1984). On occasion, boutons ascending through the INL toward the OPL were observed, similar to a rarely observed structure described in carp retina (Negishi *et al.*, 1985). Unlike Yamada *et al.* (1980) we did not observe any labeled cells in the ganglion cell layer, the so-called SG*b* type in the zebrafish retina.

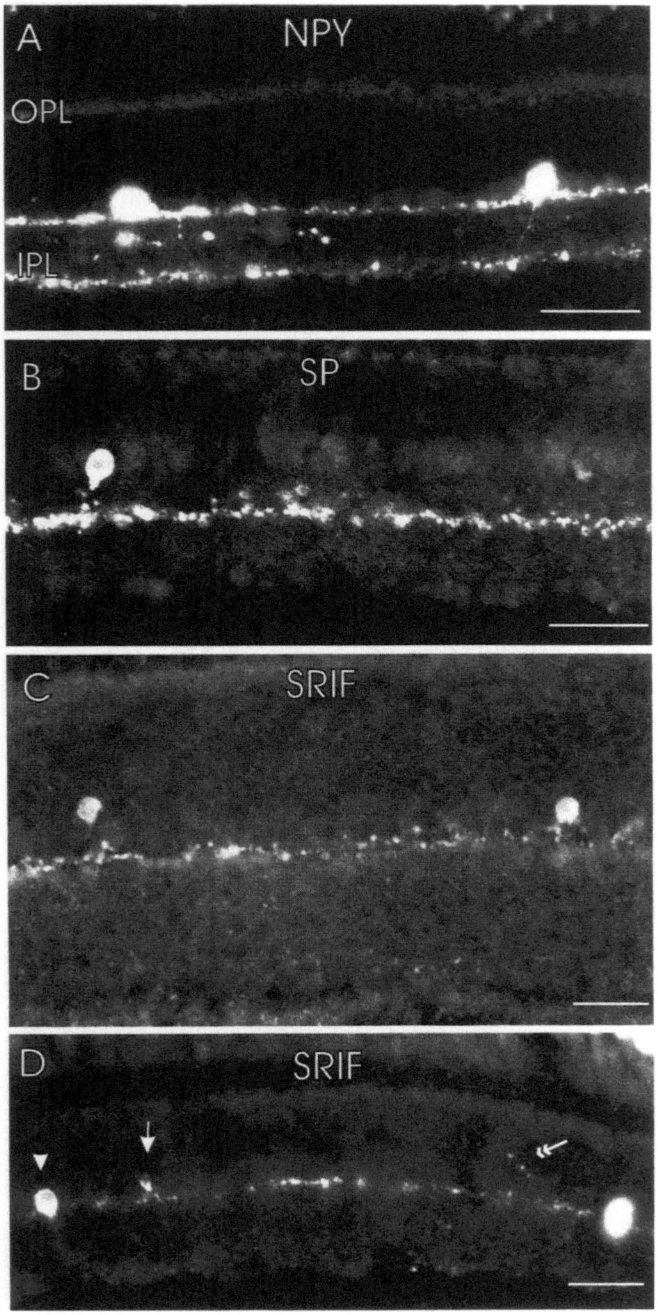

ION CHANNELS AND MEMBRANE PROTEINS

This section includes a description of voltage-gated sodium and potassium channels, a chloride transporter regulator (CFTR), connexin 26 (CX26) and synaptic vesicle protein 2 (SV2). Synaptic associated protein 97 (SAP97) did not label the zebrafish retina at all.

Sodium channels

The PanNa antiserum recognizes TTX-sensitive Na channels in all preparations so far investigated (Dugandzija-Novakovic *et al.* 1995). In zebrafish retina, PanNa-IR (Fig. 23A) was observed in the OPL, numerous cell bodies in the middle portion of the INL, three layers in the IPL (1, 3, 5) and in the optic fiber layer. Cell bodies at the proximal and distal borders of the INL were not labeled. PanNa-IR in the optic fiber layer was due to labeled axons of the ganglion cells, as numerous fibers in the optic nerve head were labeled extensively (not shown). The shape and position of the PanNa-IR cell bodies in the INL were suggestive of bipolar cells. Double labeling with PKC-IR showed only a partial overlap of label (Fig. 23A and B). ON bipolar cell bodies, labeled by PKC-IR, tend to be proximal to the OFF bipolar cell bodies. The ON mixed rod/cone type (M*b*) are a bit larger, brighter and slightly more distal to the numerous ON cone bipolar cells. The bright PanNa-IR bipolar cell bodies in the distal INL did not co-localize with PKC-IR M*b* bipolar cells, nor did cell bodies in the

Fig. 22. More Neuropeptides. Scale bars = 20 μm. (A) Neuropeptide Y—NPY-IR appeared in bistratified amacrine cells with cell bodies at the proximal margin of the INL that gave rise to two dense plexes in layers 1 and 5 of the IPL. Occasional boutons were observed in the mid IPL. The process connecting the laminas in the IPL appeared to be thinner than the processes comprising either of the main arbors. (B) Substance P—SP-IR was found in a single type of pyriform amacrine cell that sent a single process into the IPL and ramified exclusively in lamina 3. The cell bodies were rare; usually only two or three were visible in a 3 mm section of the retina. (C and D) Somatotropin releasing inhibitory factor (somatostatin)—SRIF-IR was observed in three types of amacrine cell. The SA*a*1 and SA*a*2 types each had their processes restricted to layer 1 of the IPL. However, the SA*a*1 type (C) had a larger cell body and smooth dendrites whereas the SA*a*2 type (D, arrow) had a small cell body and varicose processes. The Sa*b* type was a pyriform cell that gave rise to a thin process that ramified in layer 5 of the IPL (D, arrowhead). In addition, fine processes were occasionally observed extending from the IPL to the OPL (D, double arrowhead).

mid-INL, indicating that those PanNa-IR bipolar cells were of the OFF type. However, it appeared as if most, if not all of the PanNa-IR cell bodies in the proximal INL were also PKC-IR; but, there were numerous cells in the proximal INL that were only PKC-IR. The general conclusion is that PanNa-IR is found in all OFF bipolar cells and a subpopulation of ON cone bipolar cells that have cell bodies in the proximal region of the INL. These findings are the same as reported in goldfish retina (Zenisek *et al.*, 2001).

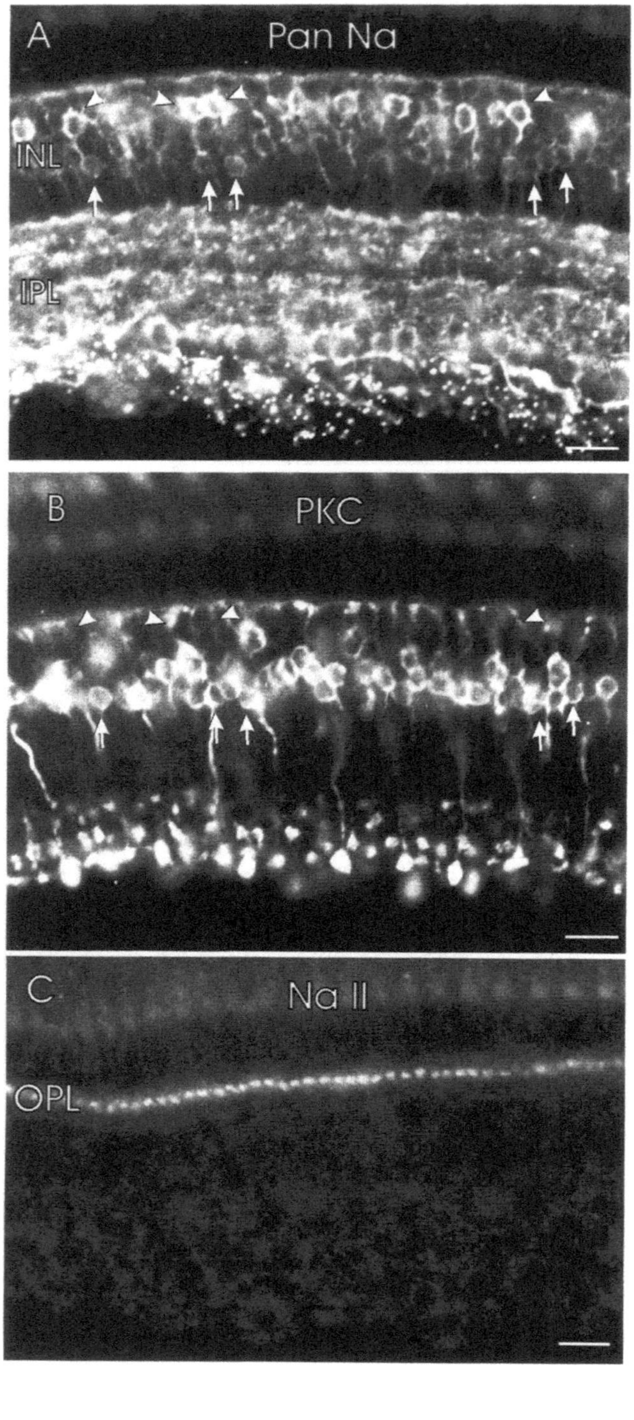

Fig. 23. Voltage-gated sodium channels. Scale bars = 20 μm. (A) PanNa-IR was observed in the OPL, numerous cell bodies in the middle portion of the INL, multiple sublayers throughout the IPL and in the optic fiber layer. Cell bodies at the proximal and distal borders of the INL were not labeled. Double labeling with PKC-IR (A and B) showed a partial overlap of the two labels. ON bipolar cell bodies, labeled by PKC-IR, tend to be proximal to OFF bipolar cell bodies. The bright PanNa-IR bipolar cell bodies in the distal INL (arrowheads) did not co-localize with PKC-IR *Mb* bipolar cells There was co-localization of the PanNa-IR cell bodies located more proximally in the INL with PKC-IR cell bodies (arrows). However, there were numerous PKC-IR cell bodies that were not double-labeled. It appears that PanNa-IR is found in OFF bipolar cells and a subpopulation of ON cone bipolar cells. (C) NaII-IR was found in a very discrete layer in the OPL, in which it appeared as a string of boutons confined to a single layer.

The NaII antiserum was generated against a unique sequence of the $Na_V1.2$ subunit C-terminus (Gong *et al.*, 1999). It labeled a very discrete layer in the OPL (Fig. 23C). In vertical section, NAII-IR appeared as regularly-spaced, bright puncta; in horizontal section, these puncta appeared as small rosettes. These structures appeared to be smaller than those that were GluR2-IR and resembled those that were GluR4-IR. This finding is peculiar in that NaII-IR should represent a portion of the PanNa-IR pattern because PanNa-IR should show all voltage gated Na channels regardless of subunit composition. However, it is clear that PanNa-IR is not very dense in the OPL, the site of intense NaII-IR. Therefore, until the NaII antiserum is characterized in the fish, conclusions about the presence of voltage gated Na-channels in the fish OPL are premature.

Potassium channels

We localized two *Shaker*-like subunits, Kv1.1 and Kv1.2, that underlay the delayed rectifier ($I_{K(v)}$) and the *Shaker*-like Kv1.4 and *Shal*-like Kv4.2 subunits that underlay the transient A-current. Unlike in goldfish (Yazulla & Studholme, 1998), we did not detect any *Shab*-like Kv2.1 subunit immunoreactivity in the zebrafish.

The Kv1.1 subunit was distributed widely in the retina (Fig. 24A). Kv1.1-IR was found in the OPL, a diffuse band in the distal INL that was likely due to horizontal cell bodies, cell bodies in the proximal INL with ascending processes that were likely to be ON cone bipolar cells, throughout the IPL with a more intense band in layer 3, and the optic fiber layer. Cell bodies in the ganglion cell layer appeared darker as they were outlined by Kv1.1-IR. The ovoid cell bodies in the INL were likely to be ON bipolar cells as they are in the same position of the PKC-IR bipolar cells. This is comparable to goldfish retina, in which Kv1.1-IR was localized to ON *Mb* and *cb* bipolar cells, as determined by

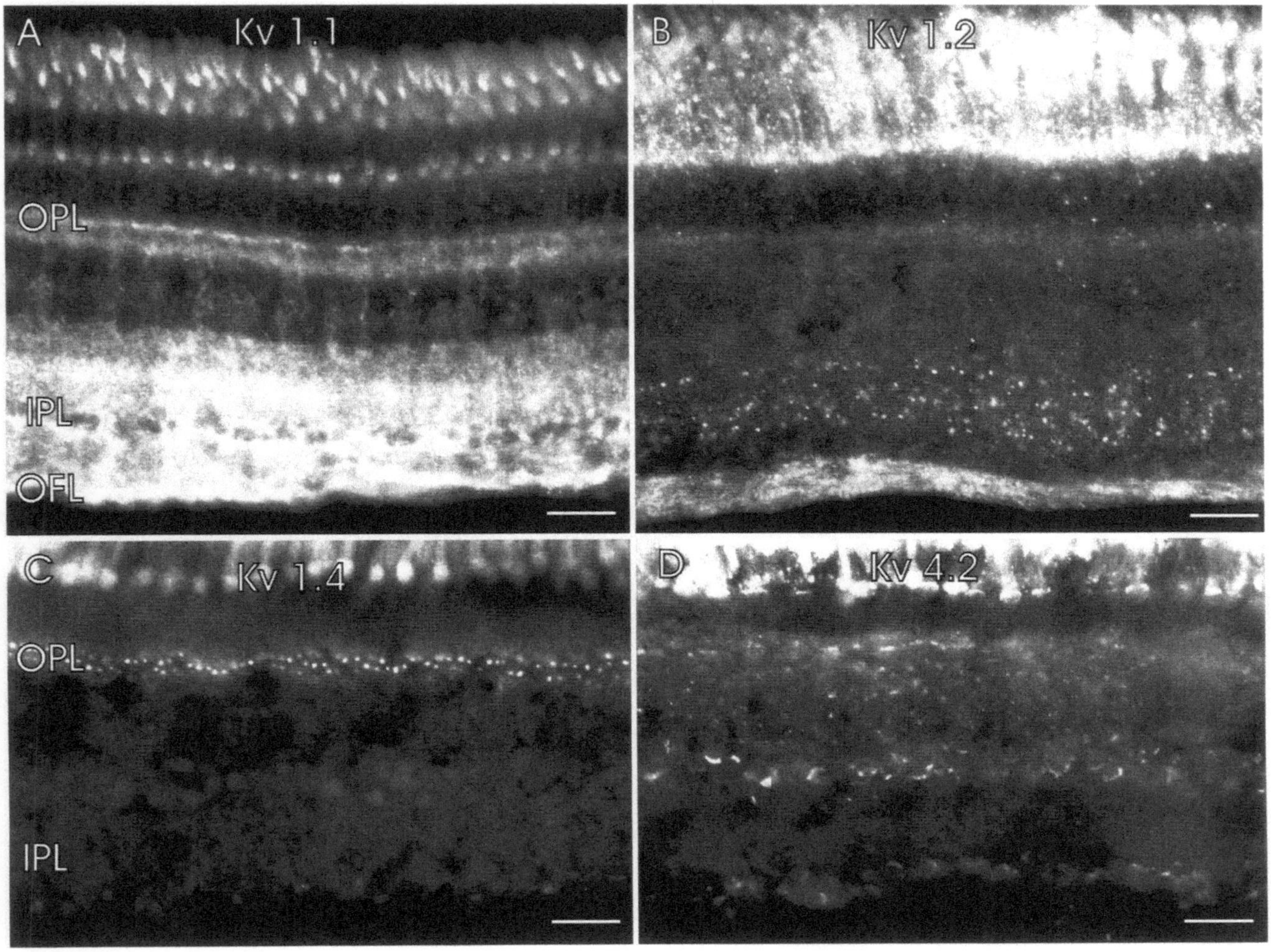

Fig. 24. Voltage-gated potassium channels. Scale bars = 20 μm. (A) Kv1.1-IR was found in the OPL, a diffuse band in the distal INL that was likely due to horizontal cell bodies, cell bodies in the proximal INL with ascending processes that were likely to be ON cone bipolar cells, throughout the IPL with a more intense band in layer 3, and the optic fiber layer. Cell bodies in the ganglion cell layer appear darker as they are outlined by Kv1.1-IR. The ovoid cell bodies in the INL are likely to be ON bipolar cells as they are in the same position of the PKC-IR bipolar cells. There also was diffuse label in the optic fiber layer. (B) There was Kv1.2-IR over the photoreceptors that appeared more punctate than the non-specific cytoplasmic "glow" seen with many of the other antibodies. There was punctate labeling in the IPL, with hints of lamination at layers 1, 3 and 4. Kv1.2-IR, in the IPL, appeared as discrete boutons rather than along fibers. This is in contrast to the optic fiber layer in which Kv1.2-IR was observed all along the optic nerve fibers. (C) The Kv.1.4 subunit was characterized by its extreme restriction to puncta narrowly distributed in the OPL, at the level of the cone pedicles. (D) Kv4.2 subunit was concentrated in more the area of cell bodies rather than in the synaptic layers. Kv4.2-IR appeared as "comma"-shaped structures at the OPL/INL border, the proximal INL and ganglion cell layer. Punctate label was present in the mid INL.

double-labeling with PKC-IR, and to the cell bodies of horizontal cells (Yazulla & Studholme, 1998). A difference was that cells in the ganglion cell layer were Kv1.1-IR in goldfish but apparently not in zebrafish. Also, there was no Kv1.1-IR in the optic fiber layer of goldfish.

The Kv1.2 subunit had a more restricted localization in the retina (Fig. 24B) than Kv1.1. There was Kv1.2-IR over the photoreceptors that appeared more punctate than the non-specific cytoplasmic "glow" seen with many of the other antibodies. Because of this different appearance we hesitate to dismiss it as non-specific. There was punctate labeling in the IPL, with hints of lamination at layers 1, 3 and 4. The Kv1.2-IR in the IPL appeared as discrete boutons rather than along fibers. This is in contrast to the optic fiber layer in which Kv1.2-IR was observed all along the optic nerve fibers. This was seen more clearly in the horizontal sections that illustrated clusters of optic fiber fascicles (not shown). Unlike in goldfish there was no evidence for Kv1.2-IR on the cell bodies or ascending dendrites of bipolar cells in zebrafish. Also, there was no Kv1.2-IR in the optic fiber layer of goldfish (Yazulla & Studholme, 1998).

The Kv1.4 subunit was characterized by its extreme restriction to puncta narrowly distributed in the OPL (Fig. 24C). This is very similar in appearance to that shown for GluR4-IR, NR2B-IR and NaII-IR (Figs. 11B, 12D and 23C). We showed in goldfish retina (Yazulla & Studholme, 1999) that Kv1.4 co-localized with GluR4 in the dendrites of OFF bipolar cells. The Kv4.2 subunit was concentrated in the area of cell bodies rather than in the synaptic layers (Fig. 24D). Kv4.2-IR appeared as "comma"-shaped structures in the OPL/INL border, the proximal INL and ganglion cell layer. Punctate label was present in the mid INL. This is similar to, but much weaker than observed in goldfish, in which the cell bodies of horizontal, amacrine cells and ganglion cells were labeled intensely at the same primary concentration (Yazulla & Studholme, 1999). The similarity is in the presence of the "comma" shaped structures in the INL and the general lack of label in the plexiform layers.

Chloride channel

The cystic fibrosis transporter regulator (CFTR) is extremely important for the movement of chloride related to osmotic regulation and is widely distributed in nature (Jentsch, 1996). CFTR-IR in the zebrafish appeared as numerous puncta distributed throughout the retina (Fig. 25A). The puncta were most intense at the vitreal surface, from which hints of streaks may be discerned in the IPL, INL and ending at a horizontal band of puncta at the outer limiting membrane. This distribution is highly suggestive of Müller's cells. There also appears to be a band of CFTR-IR in the OPL and in the region of cone horizontal cell bodies.

Postsynaptic density 95

It was difficult to obtain good labeling with either of the PSD-95 antibodies that we tried. Often the label was apparent but very weak. Light fixation and

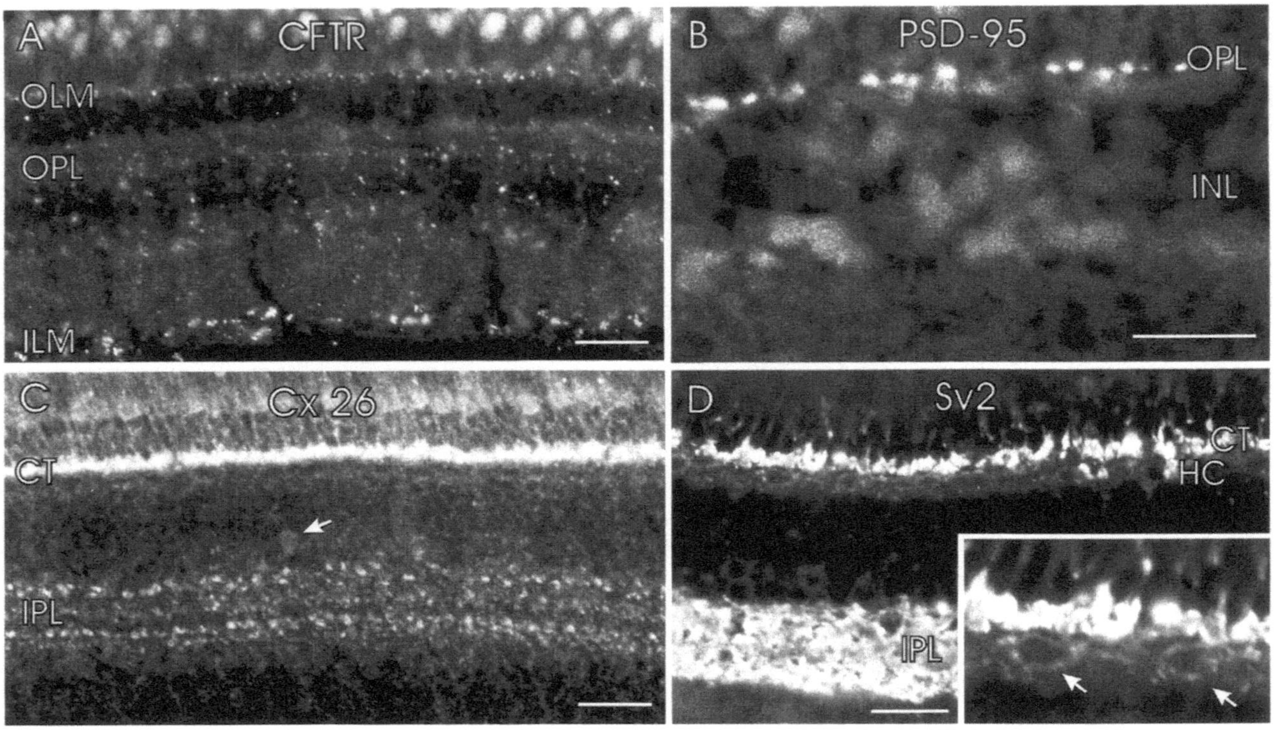

Fig. 25. Membrane and channel proteins. Scale bars = 20 μm. (A) Cystic fibrosis transport regulator—CFTR-IR appeared as numerous puncta distributed throughout the retina. The puncta were most intense at the inner limiting membrane (ILM), from which hints of streaks were observed in the IPL and INL ending at a horizontal band of puncta at the outer limiting membrane (OLM), a distribution suggestive of Müller's cells. There was a band of CFTR-IR in the OPL and region of cone horizontal cell bodies. (B) Postsynaptic density 95—PSD-95-IR was confined to a very narrow band in the OPL, at the level of the cone pedicles. (C) Connexin 26—CX26-IR appeared as a bright band in the OPL, in which it appeared associated with cone photoreceptor terminals (CT). The boutons were relatively large and appeared continuous with thin axons of cones that also were CX26-IR. Labeling in the IPL appeared as regularly-spaced puncta in two bright laminas in layers 1 and 4, with dimmer laminas in layers 2 and 5. Occasionally, CX26-IR was observed in pyriform amacrine cells (arrow). (D) Synaptic vesicle 2—SV2-IR was uniformly distributed in the OPL, in which it was associated with cone photoreceptor terminals (CT), and throughout the IPL. Boutons also were visible in the proximal and distal INL. Note, in particular the presence of SV2-IR surrounding the horizontal cell bodies, likely to be indicative of synapses made by dopamine IPCs (inset).

freshly prepared sections yielded the best results. PSD-95-IR was confined to a very narrow band in the OPL (Fig. 25B). Considering the huge number of synapses in the retina (*i.e.*, see the distribution of synaptic vesicle 2 below), the absence of PSD-95-IR in the IPL was most surprising. Identification of the PSD-95-IR structures in the OPL will be described in the Double-labeling experiments below.

Connexin-26

Gap junctions are formed by the apposition of two hemichannels (connexons), each composed of six subunits (connexins). Several connexins have been identified in the zebrafish retina by molecular techniques (Dermietzel *et al.*, 2000). One such subunit, connexin 26, has received some attention for its role in signal modulation in the goldfish retina (Kamermans *et al.*, 2001). In zebrafish, CX26-IR appeared as a bright band in the OPL (Fig. 25C), in which it appeared to be associated with photoreceptor terminals. The boutons were relatively large and appeared as a continuous band in the OPL. Labeling in the IPL was characterized by regularly-spaced puncta in two bright laminas in layers 1 and 4. One can also detect dimmer laminas in layers 2 and 5. Layer 2 was wider and probably extended to layer 3 that includes a clearly-defined dark band. Occasionally, CX26-IR was observed in a pyriform amacrine cell. There was no other obvious somatic labeling in the inner retina. This distribution of CX26-IR is nearly identical to that reported in carp (Janssen-Bienhold *et al.*, 2001). However, at the EM level, CX26-IR was localized to horizontal cell dendrites in both rods and cones in carp and goldfish (Janssen-Bienhold *et al.*, 2001; Kamermans *et al.*, 2001). Thus, the dense label in the OPL was accounted for entirely by horizontal cell dendrite labeling. This is likely to be the case for zebrafish as well. Another connexin, CX43, was localized in zebrafish retina to the RPE, outer limiting membrane, the OPL and bistratified amacrine cells with large cell bodies and large caliber processes in the IPL (Janssen-Bienhold *et al.*, 1998). However, punctate localization of CX43 required ethanol fixation; fixation in paraformaldehyde produced diffuse somatic labeling.

Synaptic vesicle 2

SV2-IR was uniformly distributed in the OPL, in which it was associated with photoreceptor terminals, and throughout the IPL. Boutons also were visible in the ganglion cell layer, proximal and distal INL (Fig. 25D). Note, in particular the presence of SV2-IR surrounding the horizontal cell bodies, likely to be indicative of synapses made by dopamine IPCs. Although SV2-IR has not been described in any other teleost retina, this distribution is consistent with the presence of synaptic contacts on the cell bodies of ganglion cells, amacrine cells and horizontal cells (Marshak & Dowling, 1987; Yazulla & Zucker, 1988; Yazulla & Studholme, 1991a, b).

INTERMEDIATE FILAMENTS

Glial fibrillary acidic protein

GFAP-IR was restricted to the proximal half of Müller's cells, in which the labeling was most intense in the end feet at the vitreal border (Fig. 26). The labeling weakened markedly as one progressed to the INL and it disappeared almost completely at the OPL. Weakly labeled cell bodies and processes extending to the OLM can be seen in the well-oriented section shown in Figure 26B. The restriction of GFAP to Müller's cells is not found in all fishes (Linser *et al.*, 1985). Intense horizontal cell labeling was reported for flounder, pinfish and mullet, whereas only Müller's cells were labeled in pipefish, killifish, only weakly in goldfish and stingray, and not at all in sturgeon. The concentration of GFAP to the proximal portion of Müller's cells is curious in light of

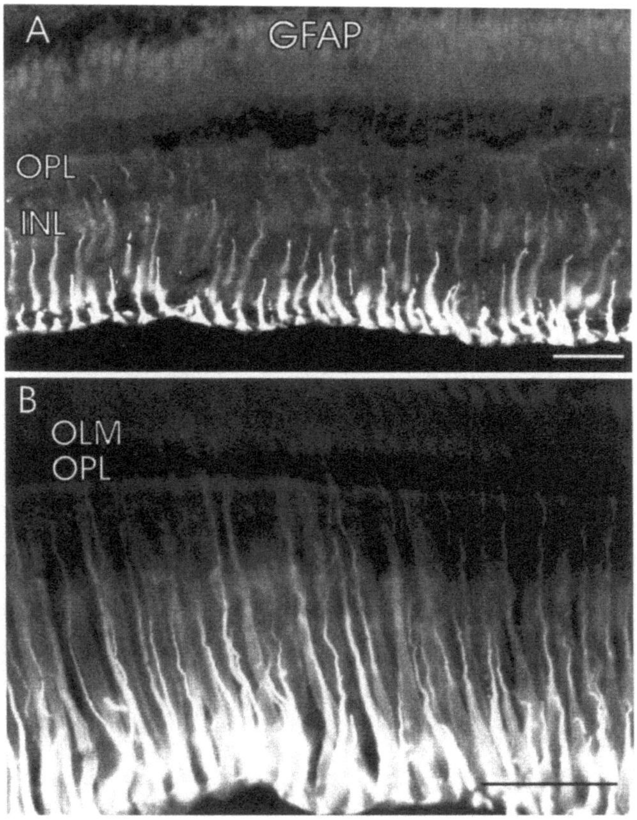

Fig. 26. GFAP-IR was restricted to the proximal half of Müller's cells, in which the labeling was most intense in the end feet at the vitreal border (A). The labeling weakened markedly as one progressed through the INL and OPL and disappeared almost completely by the level of the outer limiting membrane (OLM). Weakly labeled cell bodies and processes extending to the OPL can be seen in the well-oriented section shown in (B). Note there is virtually no label at the OLM. Scale bars = 20 μm.

the observation that both glutamine synthetase and carbonic anhydrase are uniformly distributed throughout the entire Müller's cell of zebrafish and do not show any polarization within the cell (Peterson *et al.*, 2001).

No labeling was observed with vimentin antibodies (14H7) from the Developmental Studies Hybridoma Bank. This is a bit curious because vimentin is plentiful in zebrafish lens and retina as determined by molecular techniques (Cerda *et al.*, 1998). Also, vimentin-IR has been reported in horizontal cell bodies and the proximal portion of Müller's cells in goldfish and bass (Jones & Schechter, 1987; Vaughn & Lasater, 1990), while it is restricted to Müller's cells in a cichlid fish (Mack *et al.*, 1998). This discrepancy is due almost certainly to differences in the antibodies used in each study, as 14H7 does not label the goldfish retina either.

DOUBLE-LABEL ANALYSIS OF THE OUTER PLEXIFORM LAYER

GAD67/GluR2

A number of antibodies labeled discrete structures in the OPL. As no cell bodies were labeled, the identity of these structures was unclear although it seemed likely that they represented the dendrites of horizontal cells and/or bipolar cells. Fortunately, we were able to show that GluR2-IR co-localized with GAD67-IR at the dendrites of GABAergic H1 horizontal cells (Fig. 27). Note the punctate appearance of GAD67-IR in the horizontal cell dendrites, indicating a concentration of reaction product in parts of the dendrite rather than a uniform distribution. This finding is consistent with our EM studies in goldfish showing that GluR2-IR is found in horizontal cell dendrites (Klooster *et al.*, 2001). Thus, we are able to use GluR2-IR as a marker to identify structures as H1 dendrites and by exclusion, those that were not. Those that are not, will require EM-ICC for positive identification as to the subtype of bipolar cell or as a non-GABAergic horizontal cell. The following figures, in general, make use of horizontal or oblique sections. In this view, the GluR2-IR appears as rosettes. Structures that superficially appear to co-localize in vertical sections are more easily separated in the horizontal views.

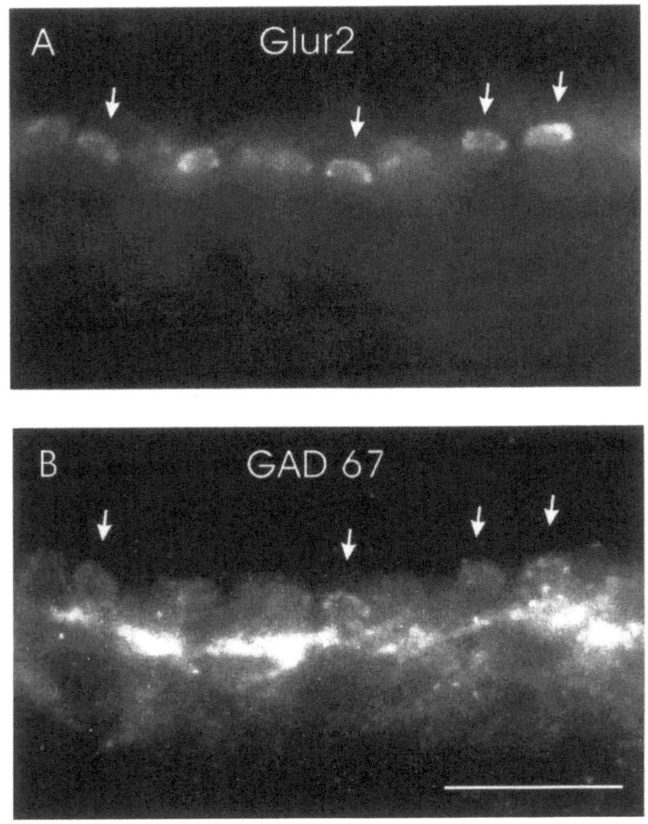

Fig. 27. Double labeling with GluR2-IR (A) and GAD67-IR (B) demonstrates that GluR2-IR is restricted to the dendrites of the GAD67-IR horizontal cells (arrows). Scale bar = 20 μm.

GluR2/GAT3

GAT3-IR was found in a thin band in the OPL in a region that appeared to include the cone terminals (see Fig. 7C). In a horizontal view, GAT3-IR appeared as a loose ring of fine puncta. Double labeling with GluR2 showed that the two labels were extensively associated with each other (Fig. 28A and B). However, as seen in a magnified overlay, the GAT3-IR appeared to surround the GluR2-IR rather than co-localize with it (Fig. 28C). This is a curious observation because GAT3, a GABA transporter, would be expected to be on the GABA accumulating horizontal cells that are labeled with GAD67. It is possible that the GAT3-IR is localized on different parts of the horizontal cell dendrite than the GluR2-IR. Again, this would require EM-ICC to corroborate.

GluR2/ADA

Our first impression of ADA-IR was that it so closely resembled GluR2-IR in terms of a rosette appearance in horizontal section that they were co-localized. However, Figure 29 shows, in vertical view, that though similar, boutons that were GluR2-IR and ADA-IR did not co-localize within the cone pedicle. Rather, the boutons appear to alternate as illustrated in the high magnification inset in Figure 29B. At this resolution, we cannot determine if the ADA-IR was on a different portion of horizontal cell dendrites or closely associated bipolar cell dendrites. The restriction of ADA-IR to dendrites innervating cones is different from that reported in goldfish in which ADA-IR was most prominent on the cell bodies, rather than dendrites, of rod and cone horizontal cells (Studholme & Yazulla, 1997). Although some puncta were visible in the goldfish OPL, it was to a far less extent than observed in the zebrafish.

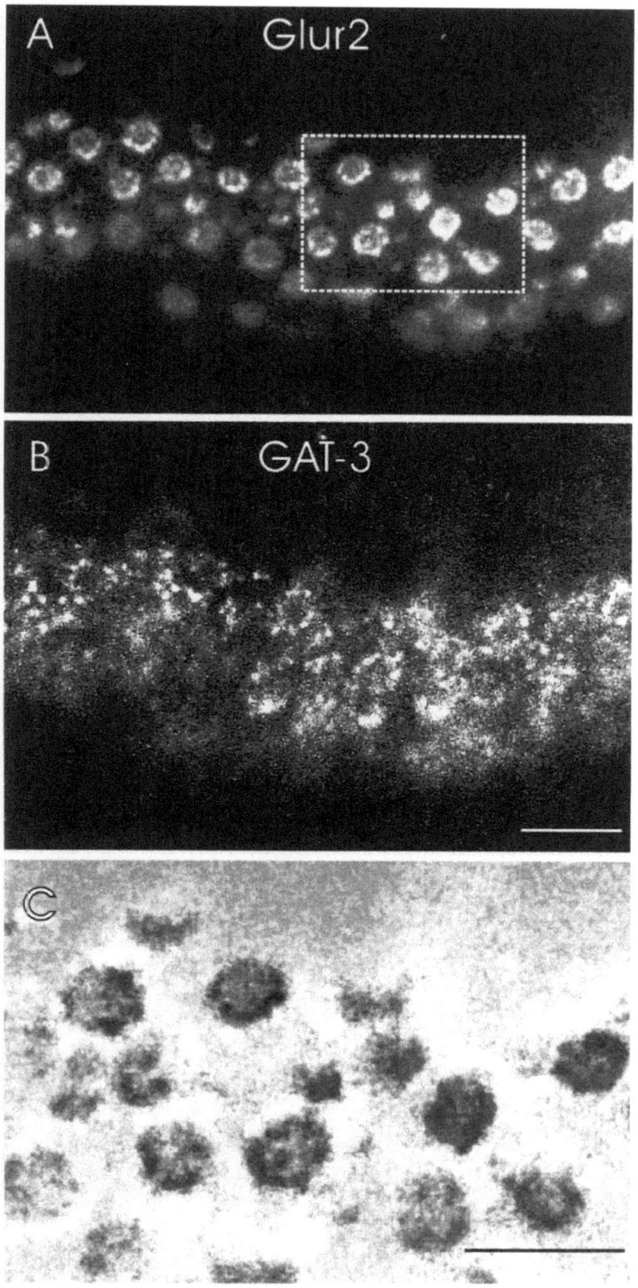

Fig. 28. Double labeling with GluR2-IR (A) and GAT3-IR (B) demonstrates that the two labels share a common pattern. The outlined area in (A) is shown in a magnified overlay with reverse contrast for GluR2-IR in (C). Here, it is clear that the dark GluR2-IR is surrounded by the light GAT3-IR. This is an oblique view of the OPL, in which the labeled dendrites appears as rosettes. Scale bar = 20 μm.

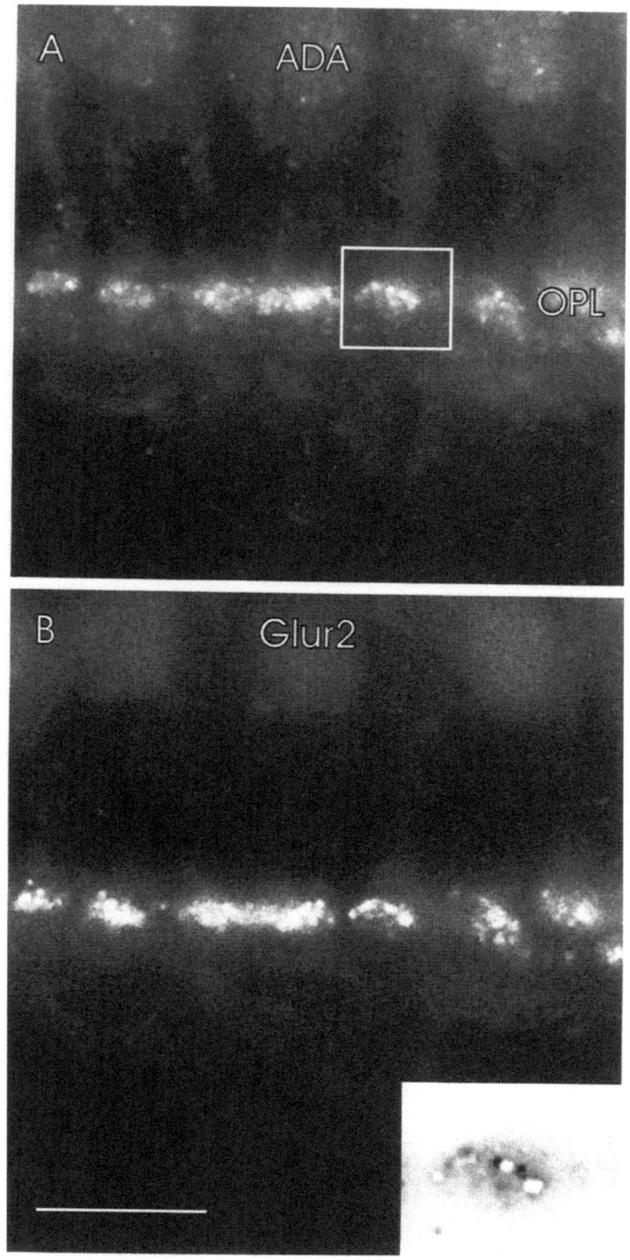

Fig. 29. Double labeling with ADA-IR (A) and GluR2-IR (B) in vertical view demonstrates that the distributions in the OPL are very similar in that each appears as boutons confined within a cone pedicle (seen as a triangular glow above the boutons). However the patterns are not identical. Inset in B) The outlined area in (A) is shown in a magnified overlay with reverse contrast for ADA-IR. Here, it is clear that the dark ADA-IR boutons alternate with the lighter GluR2-IR boutons, indicating that separate, but adjacent structures were labeled within the cone pedicle. At this resolution it cannot be determined if ADA-IR was found on the dendrites of other cone horizontal cells or bipolar cells. Scale bar = 20 μm.

GluR2/GluR4

Figure 30 clearly shows the rosette appearance of GluR2-IR that is associated with cone pedicles. Furthermore, it is clear that the GluR2-IR structures (Fig. 30B) were larger than and surrounded the GluR4-IR structures (Fig. 30A), a feature easily seen in the magnified overlay (Fig. 30C). It appeared as if the GluR4-IR fit into the GluR2-IR ring, filling gaps in the "spokes" of the rosette. Also, there were no instances in which one label was observed alone; all GluR2-IR structures were

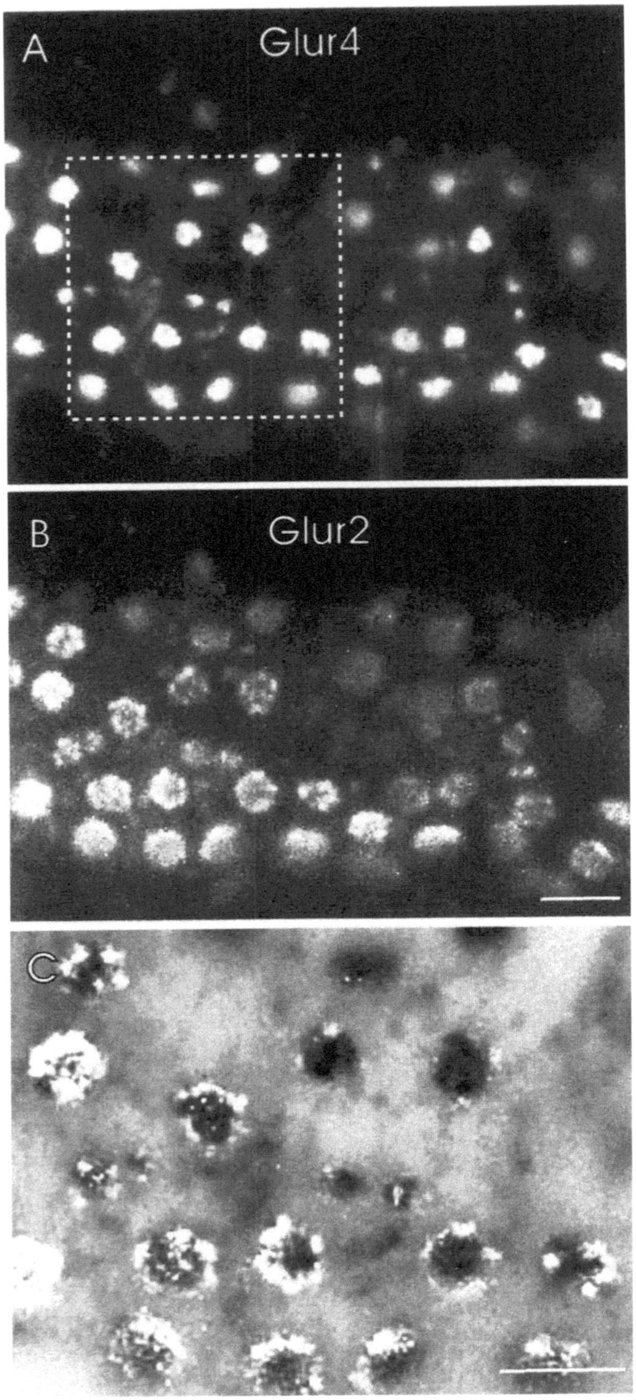

Fig. 30. Double labeling with GluR4-IR (A) and GluR2-IR (B) demonstrates that the two labels have a common pattern. The outlined area in (A) is shown in a magnified overlay with reverse contrast for GluR4-IR in (C). Here, it is clear that the dark GluR4-IR is surrounded by lighter GluR2-IR. This is an oblique view of the OPL, in which the labeled dendrites appears as rosettes. Scale bar = 20 μm.

associated with GluR4-IR structures and vice versa. This is identical to that found in goldfish, in which GluR4-IR is found on the dendrites of OFF bipolar cells (Schultz *et al.*, 1997; Yazulla *et al.*, 1999). The dendrites of OFF bipolar cells are smaller than horizontal cell dendrites and are found away from the synaptic ribbon. It is likely that this also will be found to be the case in zebrafish.

GluR2/NR2B

Every NR2B-IR structure appeared to be associated with a GluR2-IR structure (Fig. 31), similar to that observed with GluR4. The GluR2-IR structures were larger and surrounded the NR2B-IR structures as seen in the magnified overlay (Fig. 31C). The NR2B-IR structures had the same relationship with GluR2-IR as did GluR4-IR; both were surrounded. However, NR2B-IR appeared to be smaller and more concentrated in the center than GluR4-IR structures. Double labeling did not permit us to determine if NR2B-IR and GluR4-IR were on separate dendrites or on different parts of the same dendrite. The resolution to this issue will involve EM-ICC.

GluR4/PSD95

PSD95-IR was particularly faint, especially in horizontal sections. However, within the limits of resolution, it seemed clear that PSD95-IR and GluR4-IR co-localized (Fig. 32), suggesting that PSD95-IR is on the dendrites of OFF bipolar cells. This is the same result that we have obtained in goldfish (Yazulla & Studholme, unpublished observation) but different from that reported in mammalian retina, in which PSD95-IR was localized pre-synaptically in the terminals of rods and cones (Koulen *et al.*, 1998).

GluR2/GLT1

From Figure 14 there was a strong sense that GLT1-IR was localized to cone photoreceptors. The OPL was labeled intensely and the lightly labeled cone axons supported the idea that GLT1-IR was on the cone pedicles. As seen in Figure 33, the two labels have a similar rosette appearance but the shapes of the rosettes are different, suggesting that GluR2-IR and GLT1-IR were on separate, but closely associated structures, within the cone pedicle. From goldfish, we know that GLT1-IR is found in patches at the basal portion of the cone terminal and at its perimeter, not within the invaginations (VandenBranden *et al.*, 2000). GluR2-IR is found in horizontal cell dendrites, often at the periphery of the pedicle and away from the invaginations (Klooster *et al.*, 2001). It is likely that GLT1-IR will be localized to cone pedicles, as in goldfish, but EM will be required to resolve that issue.

GLT1/CB1

GLT1-IR was clearly associated with all cone terminals. It appeared, however, that CB1-IR was associated with only some cone terminals. Double labeling showed that intense CB1-IR co-localized with less than 50% of GLT1-IR terminals (Fig. 34). We were unable at this time to

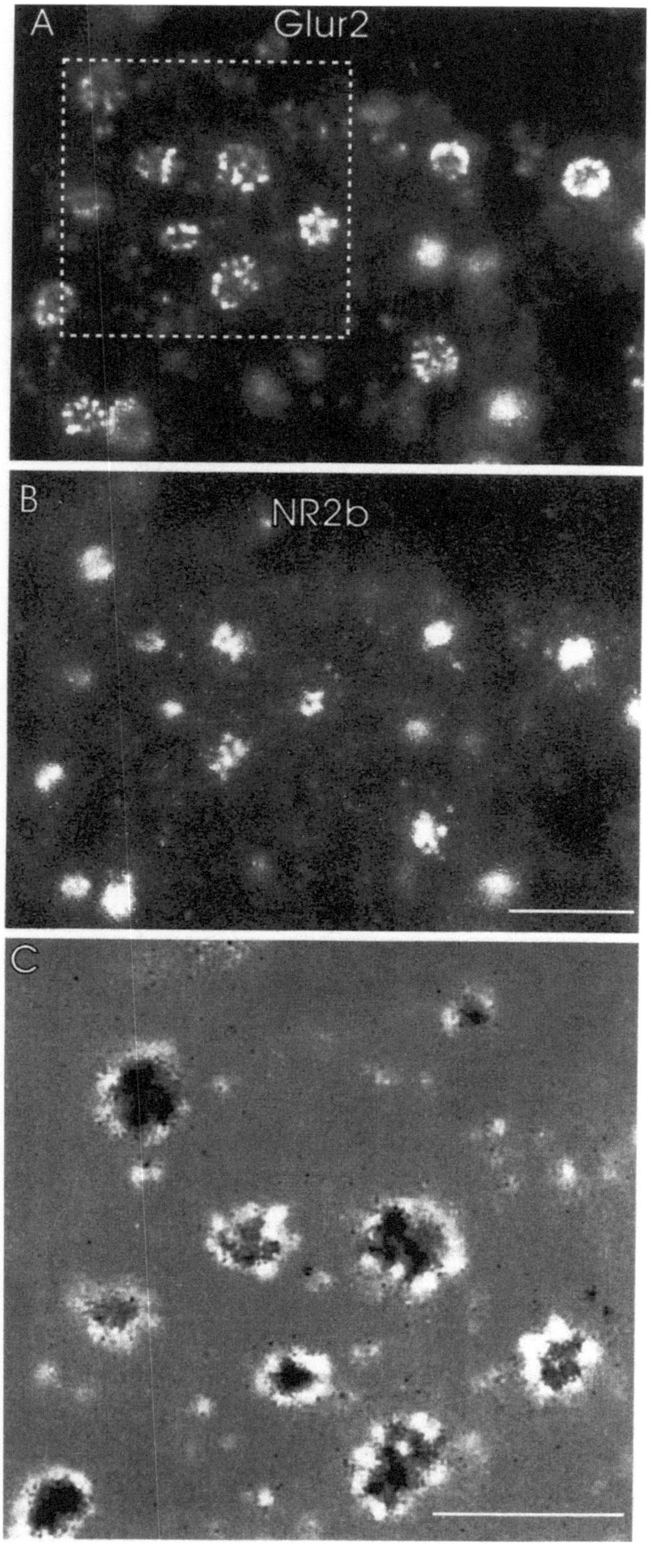

Fig. 31. Double labeling with GluR2-IR (A) and NR2B-IR (B) demonstrates that the two labels have a common pattern. The outlined area in (A) is shown in a magnified overlay with reverse contrast for NR2B-IR in (C). It is clear that the dark NR2B-IR is surrounded by lighter GluR2-IR. This is an oblique view of the OPL, in which the labeled dendrites appears as rosettes. Scale bar = 20 μm.

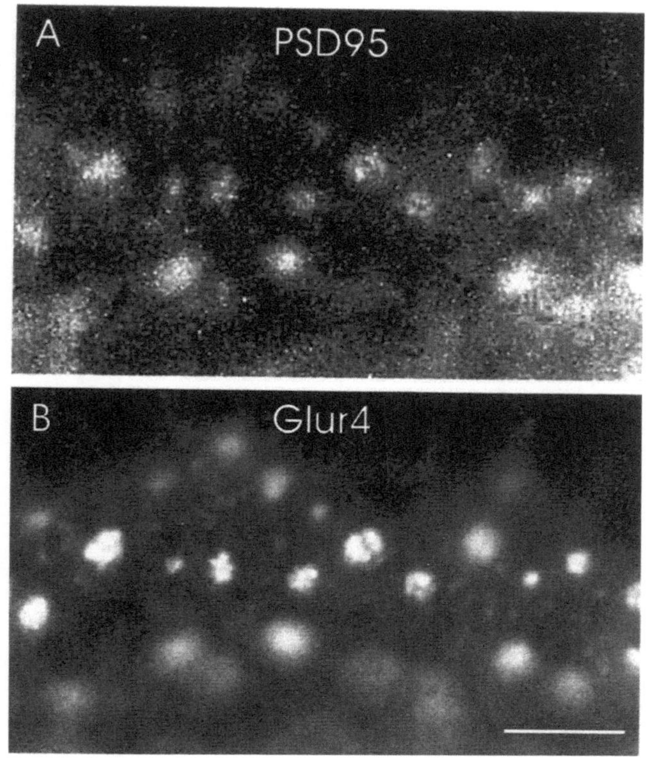

Fig. 32. Double labeling with PSD95-IR (A) and GluR4-IR (B) demonstrates an excellent correspondence in the pattern and size of the two labels indicating that these co-localize to the same structures. This is an oblique view of the OPL, in which the GluR4-IR dendrites appears as rosettes. PSD95-IR was very dim and hazy, but the overall structure and similarity to GluR4-IR can be seen. Scale bar = 20 μm.

distinguish among the cone types to determine which type was CB1-IR. This differs from goldfish in which all cone terminals appear to be CB1-IR (Yazulla *et al.*, 2000).

In summary, the double-label studies show that GluR2-IR is found on horizontal cell dendrites, whereas GluR4-IR, NR2B-IR and PSD95-IR are surrounded by GluR2-IR and, are likely to be associated with bipolar cell dendrites. The resolution of the procedure was not adequate to localize CB1-IR or GLT1-IR to cone pedicles, although data in goldfish suggest that such would be the case.

General discussion

We have described the distribution of major neurotransmitter systems in the zebrafish retina by use of light microscopic immunocytochemistry. The list of antibodies used in this study was extensive but not exhaustive. The data are summarized in Table 2, which presents an outline of the cell type and the antibody that labeled that particular cell type. As expected, when comparisons could be made, there were vast similarities between the zebrafish and goldfish. Of particular note were the

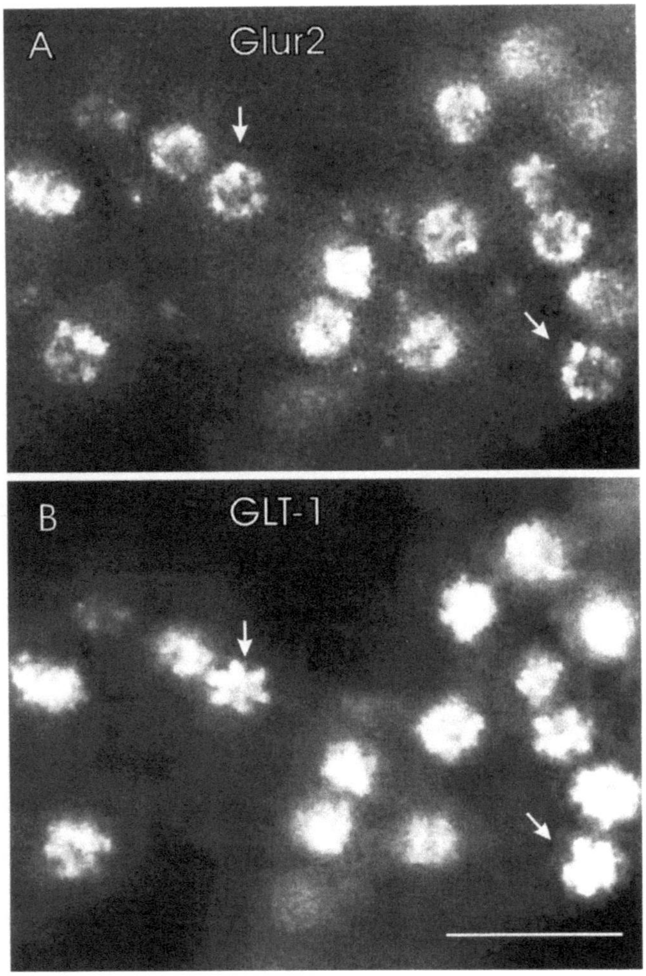

Fig. 33. Double labeling with GluR2-IR (A) and GLT1-IR (B) in a horizontal view illustrates the rosette appearance of each label, and demonstrates that the patterns, though similar, are not identical. Note the asterisk appearance of GLT1-IR as opposed to the more ring-like appearance of GluR2-IR (arrows). This suggests that GLT1-IR does not label the cone horizontal cell dendrites that are labeled by GluR2-IR. Scale bar = 20 μm.

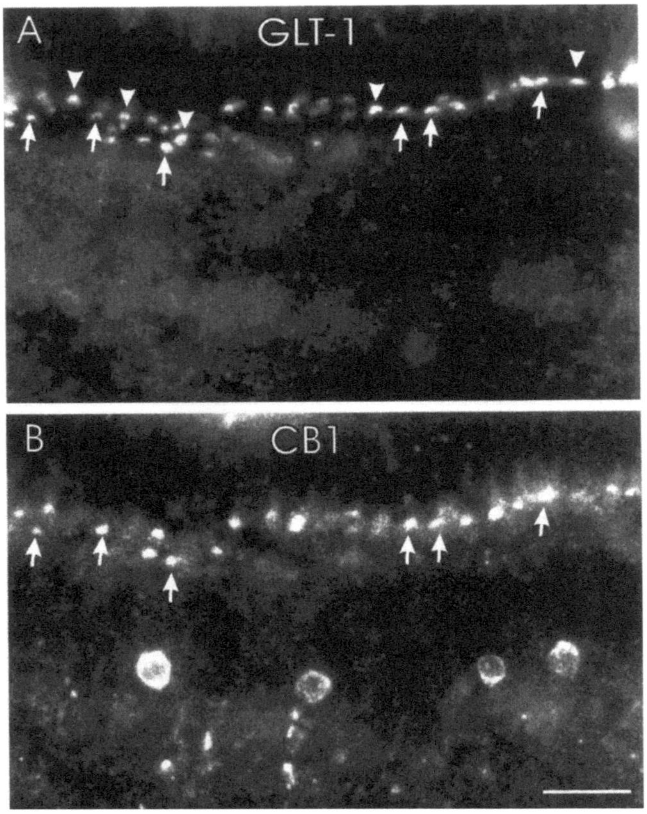

Fig. 34. Double labeling with GLT1-IR (A) and CB1-IR (B) demonstrate that all CB1-IR structures are GLT1-IR (arrows). However, about 50% of the GLT1-IR structures are not CB1-IR (arrowheads), indicating that CB1-IR is found in a subpopulation of photoreceptors. Scale bar = 20 μm.

few antibodies that label specific cell types in goldfish but not in zebrafish. These included GAD-65, glucagon, Kv2.1, neurotensin and NMDAR1. There were subtle differences with goldfish, including the GAD antisera, $GABA_A$ receptor subunits, gephyrin, PKC and the D1 receptor. There were also major differences with goldfish, including the $GABA_C\rho$ subunit, GLT-1, adenosine deaminase, the CB1 receptor and Kv1.2. These comparisons were discussed in detail as the labeling pattern of each antibody was described. Rather than reiterate these comparisons, the Discussion here will focus on some of the questions that arose and remain following the single and double label studies that were presented.

PHOTORECEPTORS

It appeared as if GluR5-7,mGluR2/3, mGlur5 and Met Enk labeled the pedicles of every cone photoreceptor. Double labeling with ZPR-1 (double cones) or recoverin (all cones) would be needed to confirm this. A more intriguing question is whether the glutamate receptor antibodies are membrane bound or intracellular. This difference would be critical because of the implication that the presence of both ionotropic and metabotropic glutamate receptors would have for photoreceptor physiology. In rat retina, the group III metabotropic glutamate subunit mGluR8 and the NMDA receptor NR1C2′ subunit were localized to rod and cone synaptic terminals (Koulen *et al.*, 1999; Fletcher *et al.*, 2000). Furthermore, group III mGluR agonists as well as glutamate reduced intracellular Ca^{2+} concentration in isolated photoreceptors, suggesting presynaptic modulation of transmitter release by autoreceptors (Koulen *et al.*, 1999). It remains to be determined if zebrafish follows this general scheme.

The granular nature of Met Enk-IR suggests an intracellular localization. The presence and significance of enkephalin in photoreceptors remains to be corroborated and determined. An effect of neuropeptides on photoreceptor function may not be so unusual. Akopian *et al.* (2000) showed that somatostatin enhanced voltage-gated K^+ currents but

Table 2. List of cell types that are labeled by specific antibodies.

Cell type	*Antibody*	*Comments*	*Figure*
Pigment epithelium	ZPR2		Figure 1A
Photoreceptors	CB1	cone pedicles	Figure 19A
	CX26	cone pedicles	Figure 25C
	Gephyrin	cone axons	Figure 7A
	GLT-1	cone pedicles	Figure 14
	GluR5–7	cone pedicles, dense	Figure 12A
	Met Enk	cone pedicles	Figure 21D
	mGluR2/3	cone pedicles	Figure 13B
	mGluR5	cone pedicles, granular	Figure 13D
	Recoverin	Diffuse—all cones	Figure 3C
	SV2	rod and cone terminals	Figure 25D
	ZNS2	Diffuse—all cones	Figure 2C
	ZPR1	Double cones	Figure 2A
Horizontal cells	ADA	HC dendrites	Figure 17B
	$GABA_A\alpha 1$	soma not axon terminal	Figure 8A
	$GABA_C\rho$	soma not axon terminal	Figure 9C
	GABA-T	distal portion of cell bodies	Figure 6
	GAD (all)	cone horizontal cell bodies	Figure 5
	GluR2	HC dendrites	Figure 11B
	GluR5–7	cell bodies	Figure 12A
	Kv1.1	weak cell bodies	Figure 24A
	mGluR2/3	punctate on cell bodies	Figure 13B
	PNMT	Thin band over distal region of soma	Figure 15B
	SV2	puncta surround HC cell bodies	Figure 25D
	VGAT	intense soma and dendrites, no axon	Figure 7D
	ZNS2	weak cell bodies	Figure 2C
Bipolar cells	D1R	Mb synaptic terminals	Figure 16A
	$GABA_A\alpha 3$	All OFF and ON Mb bipolar cells	Figure 8B
	GluR4	BC dendrites only	Figure 11D
	Kv1.4	BC dendrites only	Figure 24C
	mGluR1α	only Mb bipolar cells	Figure 13A
	NR2B	BC dendrites only	Figure 12D
	Pan Na	mix of ON and OFF BC, cone types	Figure 23A
	PKC	All ON bipolar cells	Figure 3A
	PSD-95	BC dendrites only	Figure 25B
Amacrine cells	Calretinin	large pyriform, layer 3	Figure 3B
	CB1	Few cells and processes in IPL layer 3	Figure 19A
	CCK	Few cells, processes in IPL layers 1, 4	Figure 20A
	ChAT	Large round cells, layers 1, 2, 4, 5 of IPL	Figure 18
	CX26	Few cells, boutons layers 1, 2, 4, 5	Figure 25C
	D1R	Weak cell bodies, full IPL	Figure 16A
	Enk -Leu	rare pyriform and stellate	Figure 21A
	Enk-Met	rare, pryiform bistratified	Figure 21C
	GABA-T	punctate cells, full IPL	Figure 6
	GAD (all)	numerous cells, full IPL	Figure 5
	GAT1	numerous cells, full IPL	Figure 7A
	GAT2	fewer cells, no IPL	Figure 7B
	GluR1	likely AC in proximal INL	Figure 11A
	GluR2/3	many AC, layers 2 and 4 in IPL	Figure 11C
	GluR5–7	Large AC, polarized process in layer 3	Figure 12A
	GluR6, 7	Pyriform AC, layers 1, 3, 5 in IPL	Figure 12B
	KA2	Few AC, layer 2 band	Figure 12C
	mGluR2/3	Few AC, layers 1 and 5 in IPL	Figure 13B
	NPY	Few cells, bistratified layers 1 and 5	Figure 22A
	PNMT	numerous cells, full IPL	Figure 15B
	Serotonin	Few cells, varicose processes	Figure 17A

(Continued on next page.)

Table 2. (*Continued*).

Cell type	*Antibody*	*Comments*	*Figure*
	Somatostatin	Three types, layers 1 and 5	Figure 22C
	Substance P	Few cells, monostratified in layer 3	Figure 22B
	TOH	single type of interplexiform cell	Figure 15A
	VGAT	3 IPL puncta	Figure 7D
Ganglion cell layer	Calretinin	Numerous cells, includes optic fibers	Figure 3B
	ChAT	Few small, likely to be displaced AC	Figure 18
	Enk-L	small, rare	Figure 21A
	$GABA_A\alpha3$	Numerous weakly stained cells	Figure 8B
	GAD 67	Few cells, includes some optic fibers	Figure 5D
	GAT1	Few cells—weak	Figure 7A
	GAT2	Numerous cells, probably displaced AC	Figure 7B
	GAT3	Many cells project to distal IPL	Figure 7C
	GluR2/3	Numerous small cells, displaced AC	Figure 11C
Mullers's cells	GABA-T	Possible	Figure 6
	CFTR	Punctate OLM to ILM, no soma	Figure 25A
	GFAP	Proximal portion of cell only	Figure 26
	GluR4	From OLM to ILM, weak soma	Figure 11D

suppressed voltage-gated Ca^{2+} currents in salamander photoreceptors.

From our work in goldfish, it appears as if all cone terminals contain CB1-IR (Yazulla *et al.*, 2000). This does not appear to be the case for zebrafish in which the CB1-IR is more widely spaced in the OPL, suggesting that only subtypes of cone are associated with CB1-IR. Further work is needed to determine both the cellular identity of CB1-IR localization and the subtypes of cones with which CB1 is associated.

BIPOLAR CELL BODIES

All ON bipolar cells, mixed rod/cone (M*b*) and pure cone types (C*b*), are labeled by PKC. However, only the M*b* cells are labeled by mGluR1α. The OFF bipolar cells appear to be less unequivocally labeled by any probe. The PanNa channel antibody and the $GABA_A\alpha3$ antibody appeared to label all bipolar cells except the M*b* cells. Label with these antibodies was restricted to the soma and dendrites in the OPL; axon terminals in the IPL were not labeled. Though not tested directly, we expect that there would be extensive co-localization between the PanNa channel and $GABA_A\alpha3$ subunit in C*b* cells and all OFF bipolar cells. Whether one of these will turn out to be a better marker for OFF bipolar cells remains to be determined. The absence of the $GABA_A\alpha3$ subunit and the PanNa channel from M*b* bipolar cells clearly indicates a fundamental difference in the GABA pharmacology and role of voltage-gated channels in M*b* bipolar cell function. This difference may be related to the presence of a large synaptic terminal in M*b* cells as opposed to the much smaller synaptic terminal in all other bipolar cell types.

OUTER PLEXIFORM LAYER

An issue related to the identity of OFF bipolar cells was the considerable number of antibodies that labeled small structures that were restricted to the OPL. Such structures in the OPL of goldfish have been identified by EM-ICC as belonging to cone horizontal cells—GluR2 (Klooster *et al.*, 2001) or OFF cone bipolar cells—GluR4 and Kv1.4 (Schultz *et al.*, 1997;Yazulla & Studholme, 1999). Although EM-ICC was not performed here, we showed that GluR2 co-localized with GAD67, thereby clearly localizing GluR2 to cone horizontal cell dendrites. From this identity, we were able to show that adenosine deaminase also was in cone horizontal cell dendrites. There were however, several antibodies that did not co-localize with GluR2 or with PKC (these data not shown). These antibodies included NR2B, PSD-95 and GluR4. In each case GluR2-IR surrounded these labels like a ring around a smaller ring or a core. We also showed that PSD-95 co-localized with GluR4 and, by analogy with goldfish, these may co-localize with Kv1.4 and consequently be found in OFF bipolar cell dendrites. NR2B appears even more restricted than GluR4 within the pedicle, whereas the NaII antibody appears similar in extent as GluR4 or GluR2. It thus remains to determine not only the cellular but subcellular localization of NR2B-IR and NaII-IR.

AMACRINE CELLS

Perhaps the most striking result regarding amacrine cells and the IPL was the distribution of ChAT-IR, which was found in four layers in the IPL rather than in the two layers described in every other vertebrate to date. In addition, rather than mirror symmetrical populations of orthotopic amacrine and displaced amacrine cells, the vast majority of ChAT-IR cells were orthotopic amacrine cells and they projected to all four layers. The lamination of ChAT-IR in the IPL resembled that obtained with other antibodies, particularly the glycine receptor mAb2B and $C\times 26$, in that each had four layers in what appear to be the same region. GluR2/3 had two laminas in layers 2 and 4. Double label studies would reveal potential co-localization and address the issue as to whether the ChAT-IR layers in 1 and 5 had the same glycine receptor but different glutamate receptor composition as the dendrites in layers 2 and 4.

Another potential identity is among the substance P amacrine cell, the pyriform-shaped calretinin amacrine cell and glycine amacrine cell. Although all are unistratified in sublamina 3, the arborizations and cellular densities have not been compared directly. Also, the arbor of the calretinin amacrine cell appeared denser than the substance P cell. It is possible that this difference simply reflects a differential distribution or content of antigen in the dendrites of the same cell.

Acknowledgments

We would like to acknowledge the efforts of students who participated in this project over the past several years: Undergraduates—Lev Chterenberg, Fredric Grau, David Hahn, Asna Matin, and Margaret Wong; Graduate student—Sarah Zimov. We also want to thank the numerous colleagues who were generous in supplying us with antibodies. They are listed and acknowledged in Table 1. This work was supported by NIH Grant EY01682 to Stephen Yazulla.

References

AKOPIAN, A., JOHNSON, J., GABRIEL, R., BRECHA, N. & WITKOVSKY, P. (2000) Somatostatin modulates voltage-gated K^+ and Ca^{2+} currents in rod and cone photoreceptors of the salamander retina. *Journal of Neuroscience* **20**, 929–936.

ALTSCHULER, R. A., BETZ, H., PARAKKAL, M. A., REEKS, K. A. & WENTHOLD, R. J. (1986) Identification of glycinergic synapses in the cochlear nucleus through immunocytochemical localization of the postsynaptic receptor. *Brain Research* **369**, 316–320.

BAIER, H. (2000) Zebrafish on the move: Towards a behavior-genetic analysis of vertebrate vision. *Current Opinion in Neurobiology* **10**, 451–455.

BALDRIDGE, W. H. & BALL, A. K. (1993) A new type of interplexiform cell in the goldfish retina is PNMT-immunoreactive. *Neuroreport* **4**, 1015–1018.

BEKELE-ARCURI, Z., MATOS, M. F., MANGANAS, L., STRASSLE, B. W., MONAGHAN, M. M., RHODES, K. J. & TRIMMER, J. S. (1996) Generation and characterization of subtype-specific monoclonal antibodies to K^+ channel α- and β-subunit polypeptides. *Neuropharmacology* **35**, 851–865.

BILOTTA, J. (2001) The zebrafish as a model visual system. *International Jounal of Developmental Neuroscienc* **19**, 621–629.

BLANK, H., MULLER, B. & KORF, H. (1997) Comparative investigations of the neuronal apparatus in the pineal organ and retina of the rainbow trout: Immunocytochemical demonstration of neurofilament 200-kDa and neuropeptide Y, and tracing with DiI. *Cell & Tissue Research* **288**, 417–425.

BLAZYNSKI, C. & PEREZ, M.-T. R. (1992) Neuroregulatory functions of adenosine in the retina. *Progress in Retinal Research* **11**, 293–332.

BLUTE, T. A., MAYER, B. & ELDRED, W. D. (1997) Immunocytochemical and histochemical localization of nitric oxide synthase in the turtle retina. *Visual Neuroscience* **14**, 717–729.

BOHN, M. C., DREYFUS, C. F., FRIEDMAN, W. J. & MARKIE, K. A. (1987) Glucocorticoid effects on phenylethanolamine N-methyltransferase (PNMT) in explants of embryonic rat medulla oblongata. *Developmental Brain Research* **37**, 257–266.

BRANDON, C. (1985) Retinal GABA neurons: Localization in vertebrate species using an antiserum to rabbit brain glutamate decarboxylase. *Brain Research* **344**, 286–295.

BRECHA, N. (1983) Neurotransmitters: Histochemical and biochemical studies. In *Chemical Neuroanatomy* (edited by Emson, P. C.) pp. 85–129. New York: Raven Press.

BRECHA, N., SHARMA, S. C. & KARTEN, H. J. (1981) Localization of substance P like immunoreactivity in the adult and developing goldfish retina. *Neuroscience* **6**, 2737–2746.

BRECHA, N. C. (1992) Expression of $GABA_A$ receptors in the vertebrate retina. In *Progress in Brain Research*, Vol. 90 (edited by Mize, R. R., Marc, R. E. & Sillito, A. M.) pp. 3–28. New York: Elsevier Science.

BROCKERHOFF, S. E. (2001) Retinal disease in vertebrates. In *Concepts and Challenges in Retinal Biology: A Tribute to John E. Dowling* (edited by Kolb, H., Ripps, H. & Wu, S.) Amsterdam: Elsevier; *Progress in Brain Research* **131**, 629–639.

BROCKERHOFF, S. E., HURLEY, J. B., JANSSEN-BIENHOLD, U., NEUHAUSS, S. C. F., DRIEVER, W. & DOWLING, J. E. (1995) A behavioral screen for isolating zebrafish mutants with visual system defects. *Proceedings of the National Academy of Science of the United States of America* **92**, 10545–10549.

BRUUN, A., EHINGER, B. & SYTSMA, V. M. (1984) Neurotransmitter localization in the skate retina. *Brain Research* **295**, 233–248.

CAMINOS, E., VELASCO, A., JARRIN, M., LILLO, C., JIMENO, D., AIJON, J. & LARA, J. M. (2000) A comparative study of protein kinase C-like immunoreactive cells in the retina. *Brain, Behavior & Evolution* **56**, 330–339.

CERDA, J., CONRAD, M., MARKL, J., BRAND, M. & HERRMANN, H. (1998) Zebrafish vimentin: Molecular characterization, assembly properties and developmental expression. *European Journal of Cell Biology* **77**, 175–187.

CHANG, Y.-C. & GOTTLIEB, D. I. (1988) Characterization of the proteins purified with monoclonal antibodies to glutamic acid decarboxylase. *Journal of Neuroscience* **8**, 2123–2130.

CONNAUGHTON, V. P., BEHAR, T. N., LIU, W. L. S. & MASSEY, S. C. (1999) Immunocytochemical localization of excitatory and inhibitory neurotransmitters in the zebrafish retina. *Visual Neuroscience* **16**, 483–490.

DERMIETZEL, R., KREMER, M., PAPUTSOGLU, G., STANG, A., SKERRETT, I. M., GOMES, D., SRINIVAS, M., JANSSEN-BIENHOLD, U., WEILER, R., NICHOLSON, B. J., BRUZZONE, R. & SPRAY, D. C. (2000) Molecular and functional diversity of neural connexins in the retina. *Journal of Neuroscience* **20**, 8331–8343.

DI MARZO, V. & DEUTSCH, D. G. (1998) Biochemistry of the endogenous ligands of cannabinoid receptors. *Neurobiology of Disease* **5**, 386–404.

DUGANDZIJA-NOVAKOVIC, S., KOSZOWSKI, A. G., LEVINSON, S. R. & SHRAGER, P. (1995) Clustering of Na channels and node of Ranvier formation in remyelinating axons. *Journal of Neuroscience* **15**, 492–502.

EASTER, S. S. JR. & NICOLA, G. N. (1996) The development of vision in the zebrafish (*Danio rerio*). *Developmental Biology* **180**, 646–663.

EHINGER, B. & PEREZ, M. T. R. (1984) Autoradiography of nucleoside uptake into the retina. *Neurochemistry International* **6**, 369–381.

EKMAN, R. & TORNQVIST, K. (1985) Glucagon and VIP in the retina. *Investigative Ophthalmology & Visual Science* **26**, 1405–1409.

ENZ, R., BRANDSTÄTTER, J. H., WÄSSLE, H. & BORMANN, J. (1996) Immunocytochemical localization of the $GABA_c$ receptor rho subunits in the mammalian retina. *Journal of Neuroscience* **16**, 4479–4490.

EWERT, M., DE BLAS, A. L., MÖHLER, H. & SEEBURG, P. H. (1992) A prominent epitope on $GABA_A$ receptors is recognized by two different monoclonal antibodies. *Brain Research* **569**, 57–62.

FADOOL, J. M. (2001) Understanding retinal cell fate determination through genetic manipulations. In *Concepts and Challenges in Retinal Biology: A Tribute to John E. Dowling* (edited by Kolb, H., Ripps, H. & Wu, S.) Amsterdam: Elsevier; *Progress in Brain Research* **131**, 541–554.

FAMIGLIETTI E. V. JR., KANEKO, A. & TACHIBANA, M. (1977) Neuronal architecture of on and off pathways to ganglion cells in carp retina. *Science* **198**, 1267–1269.

FLETCHER, E. L., HACK, I., BRANDSTÄTTER, J. H. & WÄSSLE, H. (2000) Synaptic localization of NMDA receptor subunits in the rat retina. *Journal of Comparative Neurology* **420**, 98–112.

FOSTER, G. A., HÖKFELT, T., COYLE, J. T. & GOLDSTEIN, M. (1985) Immunohistochemical evidence for phenylethanolamine-N-methyltransferase-positive/tyrosine hydroxylase-negative neurones in the retina and the posterior hypothalamus of the rat. *Brain Research* **330**, 183–188.

GONG, B., RHODES, K. J., BEKELE-ARCURI, Z. & TRIMMER, J. S. (1999) Type I and Type II Na+ channel alpha-subunit polypeptides exhibit distinct spatial and temporal patterning and association with auxiliary subunits in rat brain. *Journal of Comparative Neurology* **412**, 342–352.

HAMANO, K., KIYAMA, H., EMSON, P. C., MANABE, R., NAKAUCHI, M. & TOHYAMA, M. (1990) Localization of two calcium binding proteins, calbindin (28 kD) and parvalbumin (12 kD), in the vertebrate retina. *Journal of Comparative Neurology* **302**, 417–424.

HAMMANG, J. P., BOHN, M. C. & MESSING, A. (1992) Phenylethanolamine N-methyltransferase (PNMT)-expressing horizontal cells in the rat retina: A study employing double- label immunohistochemistry. *Journal of Comparative Neurology* **316**, 383–389.

HAVERKAMP, S. & WÄSSLE, H. (2000) Immunocytochemical analysis of the mouse retina. *Journal of Comparative Neurology* **424**, 1–23.

HOCH, W., BECKER, C.-M., GREENINGLOH, G. & BETZ, H. (1991) Mapping of antigenic epitopes on the α1-subunit of the inhibitory glycine receptor. *Biochemistry* **30**, 42–47.

HOLZSCHUH, J., RYU, S., ABERGER, F. & DRIEVER, W. (2001) Dopamine transporter expression distinguishes dopaminergic neurons from other catecholaminergic neurons in the developing zebrafish embryo. *Mechanisms of Development* **101**, 237–243.

HUGHES, T. E., GRÜNERT, U. & KARTEN, H. J. (1991) $GABA_A$ receptors in the retina of the cat: An immunohistochemical study of wholemounts, sections, and dissociated cells. *Visual Neuroscience* **6**, 229–238.

IMBODEN, M., DEVIGNOT, V., KORN, H. & GOBLET, C. (2001) Regional distribution of glycine receptor messenger RNA in the central nervous system of zebrafish. *Neuroscience* **103**, 811–830.

JANSSEN-BIENHOLD, U., DERMIETZEL, R. & WEILER, R. (1998) Distribution of Connexin43 immunoreactivity in the retinas of different vertebrates. *Journal of Comparative Neurology* **396**, 310–321.

JANSSEN-BIENHOLD, U., SCHULTZ, K., GELLHAUS, A., SCHMIDT, P., AMMERMÜLLER, J. & WEILER, R. (2001) Identification and localization of connexin26 within the photoreceptor-horizontal cell synaptic complex. *Visual Neuroscience* **8**, 169–178.

JENTSCH, T. J. (1996) Chloride channels: A molecular perspective. *Current Opinion in Neurobiology* **6**, 303–310.

JOHNSON, J., CHEN, T. K., RICKMAN, D. W., EVANS, C. & BRECHA, N. (1996) Multiple γ-aminobutyric acid plasma membrane transporters (GAT 1, GAT 2, GAT 3) in the rat retina. *Journal of Comparative Neurology* **375**, 212–224.

JONES, P. A. & SCHECHTER, N. (1987) Distribution of specific intermediate-filament proteins in the goldfish retina. *Journal of Comparative Neurology* **266**, 112–121.

KAMERMANS, M., FAHRENFORT, I., SCHULTZ, K., JANSSEN-BIENHOLD, U., SJOERDSMA, T. & WEILER, R. (2001) Hemichannel-mediated inhibition in the outer retina. *Science* **292**, 1178–1180.

KAUFMAN, D. L., HOUSER, C. R. & TOBIN, A. J. (1991) Two forms of the -aminobutyric acid synthetic enzyme glutamate decarboxylase have distinct intraneuronal distributions and cofactor interactions. *Journal of Neurochemistry* **56**, 720–723.

KAWAMATA, K., OHTSUKA, T. & STELL, W. K. (1990) Electron microscopic study of immunocytochemically labeled centrifugal fibers in the goldfish retina. *Journal of Comparative Neurology* **293**, 655–664.

KEYSER, K. T., KARTEN, H. J., KATZ, B. & BOHN, M. C. (1987) Catecholaminergic horizontal and amacrine cells in the ferret retina. *Journal of Neuroscience* **7**, 3996–4004.

KLOOSTER, J., STUDHOLME, K. M. & YAZULLA, S. (2001) Localization of the AMPA subunit GluR2 in the outer plexiform layer of goldfish retina. *Journal of Comparative Neurology* **441**, 155–167.

KOLB, H., LINBERG, K. A. & FISHER, S. K. (1992) Neurons of the human retina: A Golgi study. *Journal of Comparative Neurology* **318**, 147–187.

KOULEN, P., BRANDSTÄTTER, J. H., KRÖGER, S., ENZ, R., BORMANN, J. & WÄSSLE, H. (1997) Immunocytochemical localization of the $GABA_C$ receptor rho subunits in the cat, goldfish, and chicken retina. *Journal of Comparative Neurology* **380**, 520–532.

KOULEN, P., FLETCHER, E. L., CRAVEN, S. E., BREDT, D. S. & WÄSSLE, H. (1998) Immunocytochemical localization of the postsynaptic density protein PSD-55 in the mammalian retina. *Journal of Neuroscience* **18**, 10136–10149.

KOULEN, P., KUHN, R., WÄSSLE, H. & BRANDSTÄTTER, J. H. (1999) Modulation of the intracellular calcium concentration in photoreceptor terminals by a presynaptic metabotropic glutamate receptor. *Proceedings of the National Academy of Sciences of the USA* **96**, 9909–9914.

LAM, D. M. K. (1975) Synaptic chemistry of identified cells in the vertebrate retina. *Cold Spring Harbor Symposium Quantitative Biology* **40**, 571–579.

LAMBRECHT, H.-G. & KOCH, K.-W. (1991) A 26 kD calcium binding protein from bovine rod outer segments as modulator of photoreceptor guanylate cyclase. *EMBO J* **10**, 793–798.

LARISON, R. D. & BREMILLER, R. (1990) Early onset of phenotype and cell patterning in the embryonic zebrafish retina. *Development* **109**, 567–576.

LASATER, E. M., WATLING, K. J. & DOWLING, J. E. (1983) Vasoactive intestinal peptide alters membrane potential and cyclic nucleotide levels in retinal horizontal cells. *Science* **221**, 1070–1071.

LI, H., MARSHAK, D. W., DOWLING, J. E. & LAM, D. M. K. (1986) Colocalization of immunoreactive substance P and neurotensin in amacrine cells of the goldfish retina. *Brain Research* **366**, 307–313.

LI, L. (2001) Genetic and epigenetic analysis of visual system functions of zebrafish. In *Concepts and Challenges in Retinal Biology: A Tribute to John E. Dowling* (edited by Kolb, H., Ripps, H. & Wu, S.) Amsterdam: Elsevier; *Progress in Brain Research* **131**, 555–563.

LI, L. & DOWLING, J. E. (2000) Effects of dopamine depletion on visual sensitivity of zebrafish. *Journal of Neuroscience* **20**, 1893–1903.

LIN, C.-T., LI, H.-Z. & WU, J.-Y. (1983) Immunocytochemical localization of L-glutamate decarboxylase, gamma-aminobutyric acid transaminase, cysteine sulfinic acid decarboxylase, aspartate aminotransferase and somatostatin in rat retina. *Brain Research* **270** , 273–283.

LIN, Z. S. & YAZULLA, S. (1994a) Depletion of retinal dopamine increases brightness perception in goldfish. *Visual Neuroscience* **11**, 683–693.

LIN, Z. S. & YAZULLA, S. (1994b) Heterogeneity of $GABA_A$ receptors in goldfish retina. *Journal of Comparative Neurology* **345**, 429–439.

LINSER, P. J., SMITH, K. & ANGELIDES, K. (1985) A comparative analysis of glial and neuronal markers in the retina of fish: Variable character of horizontal cells. *Journal of Comparative Neurology* **237**, 264–272.

MACK, A. F., GERMER, A., JANKE, C. & REICHENBACH, A. (1998) Muller (glial) cells in the teleost retina: Consequences of continuous growth. *Glia* **22**, 306–313.

MALICKI, J. (2000) Harnessing the power of forward genetics-analysis of neuronal diversity and patterning in the zebrafish retina. *Trends in Neurosciences* **23**, 531–541.

MARC, R. E. & CAMERON, D. A. (2001) A molecular phenotype atlas of the zebrafish retina. *Journal of Neurocytology* **30**, 593–654.

MARC, R. E. & LAM, D. M. K. (1981) Glycinergic pathways in the goldfish retina. *Journal of Neuroscience* **1**, 152–165.

MARC, R. E., LIU, W.-L. S., SCHOLZ, K. & MULLER, J. F. (1988) Serotonergic and serotonin-accumulating neurons in the goldfish retina. *Journal of Neuroscience* **8**, 3427–3450.

MARSHAK, D. W. (1992) Peptidergic neurons of teleost retinas. *Visual Neuroscience* **8**, 137–144.

MARSHAK, D. W. & DOWLING, J. E. (1987) Synapses of cone horizontal cell axons in goldfish retina. *Journal of Comparative Neurology* **256**, 430–443.

MASU, M., IWAKABE, H., TAGAWA, Y., MIYOSHI, T., YAMASHITA, M., FUKADA, Y., SASAKI, H., HIROI, K., NAKAMURA, Y. & SHIGEMOTO, R. (1995) Specific deficit of the ON response in visual transmission by targeted disruption of the mGlur6 gene. *Cell* **80**, 757–765.

MATHIEU, M., TAGLIAFIERRO, G., ANGELINI, G. & VALLARINO, M. (2001) Organization of vasoactive intestinal peptide-like immunoreactive system in the brain, olfactory organ and retina of the zebrafish, *Danio rerio*, during development. *Brain Research* **888**, 235–247.

McCORMACK, C. A. & MCDONNELL, M. T. (1994) Abnormal dorsal light response in teleost fish after intraocular injection of 6-hydroxydopamine. *Journal of Fish Biology* **45**, 515–525.

McINTOSH, H. H., SONG, C. & HOWLETT, A. C. (1998) CB_1 cannabinoid receptor: Cellular regulation and distribution in N18TG2 neuroblastoma cells. *Molecular Brain Research* **53**, 163–173.

MECHOULAM, R., FRIDE, E. & DI MARZO, V. (1998) Endocannabinoids. *European Journal of Pharmacology* **359**, 1–18.

MORA-FERRER, C., YAZULLA, S., STUDHOLME, K. M. & HAAK-FRENDSCHO, M. (1999) Dopamine D1-receptor immunolocalization in goldfish retina. *Journal of Comparative Neurology* **411**, 704–714.

MUSKE, L. E., DOCKRAY, G. J., CHOHAN, K. S. & STELL, W. K. (1987) Segregation of FMRF amide-immunoreactive efferent fibers from NPY-immunoreactive amacrine cells in goldfish retina. *Cell & Tissue Research* **247**, 299–307.

NEAL, M. & CUNNINGHAM, J. (1994) Modulation by endogenous ATP of the light-evoked release of ACh from retinal cholinergic neurones. *British Journal of Pharmacology* **113**, 1085–1087.

NEAL, M. J., CUNNINGHAM, J. R. & DENT, Z. (1998) Modulation of extracellular GABA levels in the retina by activation of glial P2X-purinoceptors. *British Journal of Pharmacology* **124**, 317–322.

NEGISHI, K., KATO, S., TERANISHI, T., KIYAMA, H., KATAYAMA, Y. & TOHYAMA, M. (1985) So-called interplexiform cells immunoreactive to tyrosine hydroxylase or somatostatin in rat retina. *Brain Research* **346**, 136–140.

NEGISHI, K. & WAGNER, H. J. (1995) Differentiation of photoreceptors, glia, and neurons in the retina of the cichlid fish *Aequidens pulcher*; An immunocytochemical study. *Developmental Brain Research* **89**, 87–102.

NELSON, R., JANIS, A. T., BEHAR, T. N. & CONNAUGHTON, V. P. (2001) Physiological responses associated with kainate receptor immunoreactivity in dissociated zebrafish retinal neurons: A voltage probe study. In *Concepts and Challenges in Retinal Biology: A Tribute to John E. Dowling* (edited by Kolb, H., Ripps, H. & Wu, S.) Amsterdam: Elsevier; *Progress in Brain Research* **131**, 255–265.

NEUHAUSS, S. C., BIEHLMAIER, O., SEELIGER, M. W., DAS, T., KOHLER, K., HARRIS, W. A. & BAIER, H. (1999) Genetic disorders of vision revealed by a behavioral screen of 400 essential loci in zebrafish. *Journal of Neuroscience* **19**, 8603–8615.

NGUYEN-LEGROS, J., MARTIN-MARTINELLI, E., SIMON, A., DENOROY, L. & VIGNY, A. (1986) Colocalization of tyrosine-hydroxylase and phenylethanolamine-N-methyltransferase immunoreactivity in the rat retina: A re-examination using double labeling on semi-thin sections. *Experimental Eye Research* **43**, 575–584.

NISHIMURA, Y., SCHWARTZ, M. L. & RAKIC, P. (1986) GABA and GAD immunoreactivity of photoreceptor terminals in primate retina. *Nature (London)* **320**, 753–756.

OERTEL, W. H., SCHMECHEL, D. E., TAPPAZ, A. & KOPIN, I. J. (1981) Production of a specific antiserum to rat brain glutamic acid decarboxylase by injection of an antigen-antibody complex. *Neuroscience* **6**, 2689–2700.

OSBORNE, N. N., NICHOLAS, D. A., CUELLO, A. C. & DOCKRAY, G. J. (1981) Localization of cholecystokinin immunoreactivity in amacrine cells of the retina. *Neuroscience Letters* **26**, 31–35.

OSBORNE, N. N., NICHOLAS, D. A., DOCKRAY, G. J. & CUELLO, A. C. (1982) Cholecystokinin and substance P immunoreactivity in retinas of rats, frogs, lizards and chicks. *Experimental Eye Research* **34**, 639–649.

OSBORNE, N. N., PATEL, S., TERENGHI, G., ALLEN, J. M., POLAK, J. M. & BLOOM, S. R. (1985) Neuropeptide Y (NPY)-like immunoreactive amacrine cells in retinas of frog and goldfish. *Cell & Tissue Research* **241**, 651–656.

PATRECELLI, M. P., LASHUEL, H. A., GIANG, D. K., KELLY, J. W. & CRAVATT, B. F. (1998) Comparative characterization of a wild type and transmembrane domain-deleted fatty acid amide hydrolase: Identification of the transmembrane domain as a site for oligomerization. *Biochemistry* **37**, 15177–15187.

PENG, Y. W., BLACKSTONE, C. D. HUGANIR, R. L. & YAU, K. W. (1995) Distribution of glutamate receptor subtypes in the vertebrate retina. *Neuroscience* **66**, 483–497.

PETERSON, R. E., FADOOL, J. M., MCCLINTOCK, J. & LINSER, P. J. (2001) Muller cell differentiation in the zebrafish neural retina: Evidence of distinct early and late stages in cell maturation. *Journal of Comparative Neurology* **429**, 530–540.

PFEIFFER, F., SIMLER, R., GRENNINGLOH, G. & BETZ, H. (1984) Monoclonal antibodies and peptide mapping reveal structural similarities between the subunits of the glycine receptor of rat spinal cord. *Proceedings of the National Academy of Science of the USA* **81**, 7224–7227.

PINELLI, C., D'ANIELLO, B., SORDINO, P., MEYER, D. L., FIORENTINO, M. & RASTOGI, R. K. (2000) Comparative immunocytochemical study of FMRFamide neuronal system in the brain of *Brachydanio rerio* and *Acipenser ruthenus* during development. *Developmental Brain Research* **119**, 195–208.

RICHARDS, J. G., SCHOCH, P, HARING, P., TAKACS, B. & MOHLER, H. (1987) Resolving $GABA_A$/benzodiazepine receptors: Cellular and subcellular localization in the CNS with monoclonal antibodies. *Journal of Neuroscience* **7**, 1866–1876.

SANDELL, J. H., MARTIN, S. C. & HEINRICH, G. (1994) The development of GABA immunoreactivity in the retina of the zebrafish (*Brachydanio rerio*). *Journal of Comparative Neurology* **345**, 596–601.

SANTOS, P. F., CARAMELO, O. L., CARVALHO, A. P. & DUARTE, C. B. (2000) Adenosine A_1 receptors inhibit Ca^{2+} channels coupled to the release of ACh, but not of GABA, in cultured retina cells. *Brain Research* **852**, 10–15.

SASSOÈ-POGNETTO, M., KIRSCH, J., GRÜNERT, U., GREFERATH, U., FRITSCHY, J. M., MÖHLER, H., BETZ, H. & WÄSSLE, H. (1995) Colocalization of gephyrin and $GABA_A$-receptor subunits in the rat retina. *Journal of Comparative Neurology* **357**, 1–14.

SCHOCH, P., HÄRING, P., TAKACS, B., STÄHLI, C. & MÖHLER, H. (1984) A GABA/benzodiazepine receptor complex from bovine brain: Purification, reconstitution and immunological characterization. *Journal of Receptor Research* **4**, 189–200.

SCHULTZ, K., GOLDMAN, D. J., OHTSUKA, T., HIRANO, J., BARTON, L. & STELL, W. K. (1997) Identification and localization of an immunoreactive AMPA-type glutamate receptor subunit (GluR4) with respect to identified photoreceptor synapses in the outer plexiform layer of goldfish retina. *Journal of Neurocytology* **26**, 651–666.

SCHULTZ, K., JANSSEN-BIENHOLD, U. & WEILER, R. (2001) Selective synaptic distribution of AMPA and kainate receptor subunits in the outer plexiform layer of the carp retina. *Journal of Comparative Neurology* **435**, 433–449.

SHENG, M., TSAUR, M. L., JAN, Y. N. & JAN, L. Y. (1992) Subcellular segregation of two A-type K^+ channel proteins in rat central neurons. *Neuron* **9**, 271–284.

STELL, W. K., ISHIDA, A. T. & LIGHTFOOT, D. O. (1977) Structural basis for On- and Off-center responses in retinal bipolar cells. *Science* **198**, 1269–1271.

STELL, W. K., WALKER, S. E., CHOHAN, K. S. & BALL, A. K. (1984) The goldfish nervous terminalis: A luteinizing hormone-releasing hormone and molluscan cardioexcitatory peptide immunoreactive olfactoretinal pathway. *Proceedings of the National Academy of Science of the USA* **81**, 940–944.

STRAIKER, A., STELLA, N., PIOMELLI, D., MACKIE, K., KARTEN, H. J. & MAGUIRE, G. (1999) Cannabinoid CB1 receptors and ligands in vertebrate retina: Localization and function of an endogenous signaling system. *Proceedings of the National Academy of Sciences of the USA* **96**, 14565–14570.

STUDHOLME, K., GLASER, S., DEUTSCH, D. G. & YAZULLA, S. (2000) Retinal cannabinoids: Comparative distribution of ^{3}H-anadamide uptake and fatty acid amide hydrolase immunoreactivity in goldfish retina. *Investigative Ophthalmology & Visual Science (Suppl.)* **41**, S296.

STUDHOLME, K. M. & YAZULLA, S. (1997) ^{3}H-adenosine uptake selectively labels rod horizontal cells in goldfish retina. *Visual Neuroscience* **14**, 207–212.

SU, Y. Y. T., FRY, K. R., LAM, D. M. K. & WATT, C. B. (1986) Enkephalin in the goldfish retina. *Cellular and Molecular Neurobiology* **6**, 331–348.

SU, Y. Y. T., WU, J.-Y. & LAM, D. M. K. (1979) Purification of L-glutamic acid decarboxylase from catfish brain. *Journal of Neurochemistry* **33**, 169–179.

SUBHEDAR, N., CERDA, J. & WALLACE, R. A. (1996) Neuropeptide Y in the forebrain and retina of the killifish, *Fundulus heteroclitus*. *Cell and Tissue Research* **283**, 313–323.

SUZUKI, S. & KANEKO, A. (1990) Identification of bipolar cell subtypes by protein kinase C-like immunoreactivity in the goldfish retina. *Visual Neuroscience* **5**, 223–230.

TERANISHI, T., NEGISHI, K., HIDAKA, S. & NAKA, K. (1987) Dendritic morphology of indoleamine cells revealed by intracellular injection of Lucifer Yellow in fixed carp retina. *Neuroscience* **22**, 323–329.

TORNQVIST, K. & EHINGER, B. (1983) Glucagon immunoreactive neurons in the retina of different species. *Graefes Archive for Clinical & Experimental Ophthalmology* **220**, 1–5.

TRILLER, A., CLUZEAUD, F. & KORN, H. (1987) Gamma-aminobutyric acid-containing terminals can be apposed to glycine receptors at central synapses. *Journal of Cell Biology* **104**, 947–956.

TUMOSA, N., ECKSTEIN, F. & STELL, W. K. (1984) Immunocytochemical localization of putative cholinergic neurons in the goldfish retina. *Neuroscience Letters* **48**, 255–259.

TUMOSA, N. & STELL, W. K. (1986) Choline acetyltransferase immunoreactivity suggests that ganglion cells in the goldfish retina are not cholinergic. *Journal of Comparative Neurology* **244**, 267–275.

VANDENBRANDEN, C. A. V., KAMPHUIS, W., CARDOZO, B. N. & KAMERMANS, M. (2000b) Expression and localization of ionotropic glutamate receptor subunits in the goldfish retina-an *in situ* hybridization and immunocytochemistry study. *Journal of Neurocytology* **29**, 729–742.

VANDENBRANDEN, C. A. V., YAZULLA, S., STUDHOLME, K. M., KAMPHUIS, W. & KAMERMANS, M. (2000a) Immunocytochemical localization of the glutamate transporter GLT-1 in goldfish (*Carassius auratus*) retina. *Journal of Comparative Neurology* **423**, 440–451.

VAN HAESENDONCK, E., MARC, R. E. & MISSOTTEN, L. (1993) New aspects of dopaminergic interplexiform cell organization in the goldfish retina. *Journal of Comparative Neurology* **333**, 503–518.

VARDI, N., MASARACHIA, P. & STERLING, P. (1992) Immunoreactivity to $GABA_A$ receptor in the outer plexiform layer of the cat retina. *Journal of Comparative Neurology* **320**, 394–397.

VAUGHAN, D. K. & LASATER, E. M. (1990) Glial and neuronal markers in bass retinal horizontal and Müller cells. *Brain Research* **537**, 131–140.

VITANOVA, L., KUPENOVA, P., HAVERKAMP, S., POPOVA, E., MITOVA, L. & WÄSSLE, H. (2001) Immunocytochemical and electrophysiological characterization of GABA receptors in the frog and turtle retina. *Vision Research* **41**, 691–704.

VITORICA, J., PARK, D., CHIN, G. & DE BLAS, A. L. (1988) Monoclonal antibodies and conventional antisera to the $GABA_A$ receptor/benzodiazepine receptor/Cl^- channel complex. *Journal of Neuroscience* **8**, 615–622.

WAGNER, H.-J. & WAGNER, E. (1988) Amacrine cells of a teleost fish, the roach (*Rutilus rutilis*): A Golgi study on differentiation and layering. *Philosophical Transactions of the Royal Society of London, B* **321**, 263–324.

WAGNER, H.-J. & ZEUTZIUS, I. (1987) Amacrine cells with neurotensin- and somatostatin-like immunoreactivity in three species of teleosts with different colour vision. *Cell and Tissue Research* **248**, 663–673.

WATLING, K. J. & DOWLING, J. E. (1983) Effects of vasoactive intestinal peptide and other peptides on cyclic AMP accumulation in intact pieces and isolated horizontal cells of the teleost retina. *Journal of Neurochemistry* **41**, 1205–1213.

WENTHOLD, R. J., ZEMPLE, J., PARAKKAL, M. A., REEKS, K. A. & ALTSCHULER, R. A. (1986) Immunocytochemical localization of GABA in the cochlear nucleus of the guinea pig. *Brain Research* **380**, 7–18.

YAMADA, T., MARSHAK, D., BASINGER, S., WALSH, J., MORLEY, J. & STELL, W. K. (1980) Somatostatin-like immunoreactivity in the retina. *Proceedings of the National Academy of Science of the USA* **3**, 1691–1695.

YAZULLA, S. (1986) GABAergic mechanisms in the retina. *Progress in Retinal Research* **5**, 1–52.

YAZULLA, S. (1991) The mismatch problem for GABAergic amacrine cells in goldfish retina: Resolution and other issues. *Neurochemical Research* **16**, 327–339.

YAZULLA, S., MOSINGER, J. & ZUCKER, C. (1984) Two types of pyriform Ab amacrine cells in the goldfish retina: An EM analysis of ^{3}H-GABA uptake and somatostatin-like immunoreactivity. *Brain Research* **321**, 352–356.

YAZULLA, S. & STUDHOLME, K. M. (1991a) Glycine-receptor immunoreactivity in retinal bipolar cells is postsynaptic to glycinergic and GABAergic amacrine cell synapses. *Journal of Comparative Neurology* **310**, 11–20.

YAZULLA, S. & STUDHOLME, K. M. (1991b) Glycinergic interplexiform cells make synaptic contact with amacrine cell bodies in goldfish retina. *Journal of Comparative Neurology* **310**, 1–10.

YAZULLA, S. & STUDHOLME, K. M. (1995) Volume transmission of dopamine may modulate light adaptive plasticity of horizontal cell dendrites in the recovery phase following dopamine depletion in goldfish retina. *Visual Neuroscience* **12**, 827–836.

YAZULLA, S. & STUDHOLME, K. M. (1997) Differential reinnervation of retinal bipolar cell dendrites and axon terminals by dopamine interplexiform cells following dopamine depletion with 6-OHDA. *Journal of Comparative Neurology* **382**, 535–545.

YAZULLA, S. & STUDHOLME, K. M. (1998) Differential distribution of *Shaker*-like and *Shab*-like K^+-channel subunits in goldfish retina and retinal bipolar cells. *Journal of Comparative Neurology* **396**, 131–140.

YAZULLA, S. & STUDHOLME, K. M. (1999) Co-localization of *Shaker* A-type K^+channel (Kv1.4) and AMPA-glutamate receptor (GluR4) immunoreactivities to dendrites of OFF bipolar cells of goldfish retina. *Journal of Neurocytology* **28**, 63–73.

YAZULLA, S., STUDHOLME, K. M., FAN, S. F. & MORA-FERRER, C. (2001) Neuromodulation of voltage-dependent K^+ channels in bipolar cells: Immunocytochemical and electrophysiological studies. In *Concepts and Challenges in Retinal Biology: A Tribute to John E. Dowling* (edited by Kolb, H., Ripps, H. & Wu, S.) pp. 201–214. Amsterdam: Elsevier.

YAZULLA, S., STUDHOLME, K. M., MCINTOSH, H. H. & DEUTSCH, D. G. (1999) Immunocytochemical localization of cannabinoid CB1 receptor and fatty acid amide hydrolase in rat retina. *Journal of Comparative Neurology* **415**, 80–90.

YAZULLA, S., STUDHOLME, K. M., MCINTOSH, H. H. & FAN, S. F. (2000) Cannabinoid receptors on goldfish retinal bipolar cells: Electron-microscope immunocytochemistry and whole-cell recordings. *Visual Neuroscience* **17**, 391–401.

YAZULLA, S., STUDHOLME, K. M. & PINTO, L. H. (1997) Differences in the retinal GABA system among control, *spastic* mutant and *retinal degeneration* mutant mice. *Vision Research* **37**, 3471–3482.

YAZULLA, S., STUDHOLME, K. M., VITORICA, J. & DE BLAS, A. L. (1989) Immunocytochemical localization of $GABA_A$ receptors in goldfish and chicken retinas. *Journal of Comparative Neurology* **280**, 15–26.

YAZULLA, S., STUDHOLME, K. M. & WU, J.-Y. (1986) Comparative distribution of ^{3}H-GABA uptake and GAD immunoreactivity in goldfish retinal amacrine cells: A double- label analysis. *Journal of Comparative Neurology* **244**, 149–162.

YAZULLA, S., STUDHOLME, K. M. & ZUCKER, C. L. (1985) Synaptic organization of substance P-like immunoreactive amacrine cells in goldfish retina. *Journal of Comparative Neurology* **231**, 232–238.

YAZULLA, S. & ZUCKER, C. L. (1988) Synaptic organization of dopaminergic interplexiform cells in the goldfish retina. *Visual Neuroscience* **1**, 13–30.

YOSHIDA, K., WATANABE, D., ISHIKANE, H., TACHIBANA, M., PASTAN, I. & NAKANISHI, S. (2001) A key role of starburst amacrine cells in originating retinal directional selectivity and optokinetic eye movement. *Neuron* **30**, 771–780.

YOUNG, L. H. Y. & DOWLING, J. E. (1989) Localization of cyclic adenosine monophosphate in the teleost retina: Effects of dopamine and prolonged darkness. *Brain Research* **504**, 57–63.

ZENISEK, D., HENRY, D., STUDHOLME, K. M., YAZULLA, S. & MATTHEWS, G. (2001) Voltage-dependent sodium channels expressed in nonspiking retinal bipolar cells. *Journal of Neuroscience* **21**, 4543–4550.

ZUCKER, C., YAZULLA, S. & WU, J. Y. (1984) Non-correspondence of ^{3}H-GABA uptake and GAD localization in goldfish amacrine cells. *Brain Research* **298**, 154–158.

Journal of Neurocytology **30**, 593–654 (2001)

A molecular phenotype atlas of the zebrafish retina

ROBERT E. MARC[1,*] and DAVID CAMERON[2]

[1] *John Moran Eye Center, Department of Ophthalmology, University of Utah School of Medicine, Salt Lake City, UT 84132, USA;*
[2] *Department of Neuroscience and Physiology, SUNY Upstate Medical University, Syracuse NY 13210, USA*
robert.marc@hsc.utah.edu

Received 14 December 2001; revised and accepted 25 February 2002

Abstract

The rasborine cyprinid *Danio rerio* (the zebrafish) has become a popular model of retinal function and development. Its value depends, in part, on validation of homologies with retinal cell populations of cyprinine cyprinids. This atlas provides raw and interpreted molecular phenotype data derived from computationally classified sets of small molecule signals from different cell types in the zebrafish retina: L-alanine, L-aspartate, L-glutamine, L-glutamate, glutathione, glycine, taurine and γ-aminobutyrate. This *basis set* yields an 8-dimensional signature for every retinal cell and formally establishes molecular signature homologies with retinal neurons, glia, epithelia and endothelia of other cyprinids. Zebrafish photoreceptor classes have been characterized previously: we now show their metabolic profiles to be identical to those of the corresponding photoreceptors in goldfish. The inner nuclear layer is partitioned into precise horizontal, bipolar and amacrine cell layers. The horizontal cell layer contains at least three and perhaps all four known classes of cyprinine horizontal cells. Homologues of cyprinid glutamatergic ON-center and OFF-center mixed rod-cone bipolar cells are present and it appears likely that all five classes are present in zebrafish. The cone bipolar cells defy simple analysis but comprise the largest fraction of bipolar cells, as in all cyprinids. Signature analysis reveals six molecular phenotypes in the bipolar cell cohort: most are superclasses. The amacrine cell layer is composed of $\approx$64% GABA+ and 35% glycine+ amacrine cells, with the remainder being sparse dopaminergic interplexiform cells and other rare unidentified neurons. These different amacrine cell types are completely distinct in the dark adapted retina, but light adapted retinas display weak leakage of GABA signals into many glycinergic amacrine cells, suggesting widespread heterocellular coupling. The composition of the zebrafish ganglion cell layer is metabolically indistinguishable from that in other cyprinids, and the signatures of glial and non-neuronal cells display strong homologies with those in mammals. As in most vertebrates, zebrafish Müller cells possess a high glutamine, low glutamate signature and contain the dominant pool of glutathione in the neural retina. The retinal pigmented epithelium shows a general mammalian signature but also has exceptional glutathione content (5–10 mM), perhaps required by the unusually high oxygen tensions of teleost retinas. The optic nerve and the marginal zone of the retina reveal characteristic metabolic specializations. The marginal zone is strongly laminated and its nascent neurons display their characteristic signatures before taking their place in the retina proper.

Introduction

This atlas surveys the comparative molecular neuroanatomy of the zebrafish using computational molecular phenotyping methods. The description of the structural neurochemistry of the retina of the teleost cyprinid (*Danio rerio*) primarily focuses on cellular and chemical homologies with other cyprinid fishes and, where possible, with other vertebrates including mammals. Though the cyprinid classification might seem taxonomically narrow, alas, it is not. Over 2000 extant cyprinid species are split into >200 genera, with successful invasions of virtually every freshwater niche. New species are described yearly. Within this complex, zebrafish are the only members of the small subfamily Rasborinae to come into common scientific use as a model. Various members of the more diverse and cosmopolitan subfamily Cyprininae have served as physiological models, but are poor genetic tools. The need for a decisive comparison seems clear. Formal cellular homology has been a knotty problem in neuroscience, and the task has largely depended on *ad hoc* comparisons of structure. We here employ comprehensive molecular phenotyping based on signature groupings of small molecules invariant across taxa (Marc *et al.*, 1995).

The diverse photic histories of vertebrates impact our abilities to establish vertebrate homologies among extant taxa. Teleost fishes are of relatively modern emergence, expanding first in the early Cenozoic and leaving their legacy in the teeming Eocene fossil beds of Italy and the western United States. A last common ancestor with the tetrapoda predated the

* To whom correspondence should be addressed.

0300–4864

Osteichthyan-Sarcopterygian divergence and teleosteans are now the most diverse extant vertebrate group and evince the most extensive displays of retinal specializations known, many of which may be apomorphies. Teleosts have undergone ≈400 million years of uninterrupted selection, while mammals experienced a cataclysmic reduction in retinal complexity during their emergence, most likely in the early Jurassic. Early mammals collapsed in size from their large mammal-like reptilian predecessors to tiny insectivores resembling modern shrews and voles (Carroll, 1988). Along with this new body plan, early mammals were nocturnal, fossorial and cryptic: environments biased for myopic, achromatic and largely scotopic vision. The chromatic vision of modern teleosts is highly evolved and supported by over 110 classes of retinal neurons, while that of primates is barely 20 million years in the making, with likely no more than 60 classes of retinal neurons in hand. The zebrafish retina undoubtedly contains cell types, specializations and processes alien to mammalian vision.

This atlas is not a review of zebrafish literature. It is a molecular survey of the zebrafish retina with the goal of defining retinal cell types and establishing their homologies in other vertebrates. It provides a comprehensive mapping of the zebrafish retina in different areas, orientations, scales and adaptive states. The figures are available on CD at screen resolution so data may be viewed scales higher than print permits. Our emphasis is cell classification and retinal lamination, for from these structural data we find links to physiological attributes. The retinas of cyprinines (goldfish, *Carassius auratus*; koi, *Cyprinus carpio*; common carp, *Carassius carassius*; Japanese dace, *Tribolodon hakio*; rudd, *Scardinius erythrophthlamus*; roach, *Rutilis rutilis*; tench, *Tinca tinca*) have provided the broad physiological outlines of the vision of fishes, with strong homologies to and functional analogies with the retinas of anurans, reptiles and avians (Marc, 1992). Cyprinine retinas are complex, with up to seven cone cone classes showing distinct morphologies, four horizontal cell classes, eighteen bipolar cell classes, perhaps seventy amacrine cell classes and no less than fifteen ganglion cell classes: >110 classes of neurons (Marc, 1999b). The retina of the goldfish has been extensively mapped, structurally and neurochemically (Marc *et al.*, 1978; Marc & Lam, 1981; Yazulla, 1981; Marshak *et al.*, 1982; Yazulla *et al.*, 1984; Mosinger & Yazulla, 1985; Yazulla *et al.*, 1985; Mosinger *et al.*, 1986; Yazulla *et al.*, 1986; Yazulla *et al.*, 1987; Studholme & Yazulla, 1988; Yazulla & Zucker, 1988; Kalloniatis & Marc, 1990; Marc *et al.*, 1990; Yazulla & Studholme, 1990; Yazulla, 1991; Marc, 1992; Marc *et al.*, 1995) with predominant emphasis on fast glutamatergic, GABAergic and glycinergic pathways. Is the zebrafish a goldfish? Can we circumnavigate cumbersome and challenging physiological and structural analyses in zebrafish by structural/chemical homologies that are known surrogate markers for function? We will show here that the zebrafish follows the known cyprinid *bauplan*: it is accurately patterned, strongly metabolically differentiated, and laminated at both the cellular and synaptic levels in the same manner as in other vertebrates.

The second theme of this atlas is the use of computational molecular phenotyping as a tool to differentiate cell types. Based on small molecule signatures at the single cell level, molecular phenotyping provides a compact biochemical description of cells, regardless of size, distribution, tissue or species. Though conceived and developed for the nervous system, the technology is applicable to all cells in all tissues, in all kingdoms. The most striking discovery emerging from computational molecular phenotyping is that virtually all cells, not just neurons, have metabolic specializations and useful class signatures. Of all neural complexes, the vertebrate retina is the most strongly specialized in cell types and metabolic differentiation. Computational molecular phenotyping extracts quantitative small molecule signals from every cell in a sample tissue array (Marc *et al.*, 1995).

We here use eight key molecules—a well-tested *basis set*—for retinal comparisons across taxa: L-alanine (A, ala); γ-aminobutyrate (γ, GABA); L-aspartate (D, asp); L-glutamine (Q, gln); L-glutamate (E, glu); glutathione (J, GTH); and taurine (τ, tau). The intracellular contents of these molecules reflect stoichiometric demands on source pools for (1) group transfer, (2) direct synthesis of other molecules, (3) osmoregulation, (4) redox regulation and (5) intercellular communication. There is a finite carbon skeleton budget for all cells and a fraction is dedicated to amines. The scale of this budget is millimolar, ≈10–100× greater than the levels of individual amino acids required to charge aminoacyl tRNA for protein synthesis. This makes a great variety of mixtures possible and these mixtures are characteristic signatures for neuronal classification.

Four technologies drive computational molecular phenotyping: (1) generation of serial thin section tissue arrays; (2) quantitative, reproducible immunochemistry; (3) quantitative computational signal capture and image registration; and (4) computational classification. With these tools, it becomes possible to extract a multidimensional signature from each cell in a tissue; the signature is the key to molecular phenotyping. Variations in small molecule content across cells permit statistical differentiation of classes (Marc *et al.*, 1995; Kalloniatis *et al.*, 1996; Kalloniatis & Tomisich, 1999; Marc & Jones, 2002) and provide snapshots of the metabolic functioning of each cell. While it is traditional to focus on the three amino acids glutamate, glycine and GABA as the fast neurotransmitter triumvirate of the brain and retina, we must stress that the relationships between these molecules and neurotransmitter function are not binary... not a question of presence or absence... but rather quantitative questions of sufficiency and source. How much glutamate does a photoreceptor need? How much is required for synaptic transmission? Do some signals arise due to coupling or uptake rather than synthesis? Molecular phenotyping can answer many such questions directly and set boundary conditions for others.

Computational molecular phenotyping of the zebrafish retina reveals signatures that differentiate diverse cell cohorts: the retinal pigmented epithelium (RPE), the pigmented and non-pigmented ciliary epithelium (PCE and NPCE), choroidal melanocytes and vascular cells; glial cells; and, of course, many retinal neurons. The signatures of virtually all cell types strongly match those established for mammalian counterparts, including the epithelial variants. This provides an strong argument for using the zebrafish as a model of general ocular processes such as fluid transport, intraocular pressure regulation, and uveal tract functions. The general similarity of zebrafish and primate Müller cell signatures suggests that zebrafish-based models of glutamate regulation should be valid for all vertebrates. New neurons are continually added to the margin of the zebrafish retina during growth in the adult, and it can regenerate retinal neurons after retinal damage or ablation (Cameron, 2000; Cameron & Carney, 2000; Cameron & Powers, 2000; Stenkamp *et al.*, 2001). The molecular signatures of cells in the marginal zone are already laminated, prefiguring the emergence of photoreceptors, inner nuclear layer neurons and ganglion cell layer neurons.

Materials and methods

CLASS TERMINOLOGY

Retinas are built from six formal cell *types*, distinguished by position and shape: photoreceptor, bipolar, horizontal, amacrine, ganglion and glial cells. But each type admits division based on morphology, connectivity and neurochemistry. In computational analysis, we define classes and superclasses as subsets of a type extracted by some quantitative method such as molecular phenotyping. A statistical class is defined by its position and dispersion in N-dimensional signal space and if it is well separated from its fellows and cannot be separated further in additional dimensions of any nature, it may be a *natural class*: a true homogeneous, biological cell class. That conclusion is strengthened if the statistical class possesses additional features associated with the natural classes. Alternatively, groups of natural classes may share so many molecular features that they can be grouped as a statistical *superclass*. In this atlas we will describe subsets of cell types that are likely to represent natural classes (statistical classes) and subsets that are clearly assemblies of several natural classes (statistical superclasses). Further details are given in Marc and Jones (2002). For conciseness, we sometimes denote the small molecule signals and their enrichment or depletion as single letter codes. Thus glutathione-rich, glutamate-poor cells can be termed J+ E−.

STRUCTURAL TERMINOLOGY

The cell types and classes that comprise the known cyprinid retinas have been formally named and we apply those conventions to the zebrafish in detail. Cyprinid cones fall into morphologically distinct classes (Marc, 1982, 1999b), which each contain a specific dominant pigment and engage in cone-specific synaptic connectivity patterns (Scholes & Morris, 1973; Scholes, 1975; Stell & Harosi, 1975; Marc & Sperling, 1976, 1977). Zebrafish cones are similar to those of other cyprinids (Raymond *et al.*, 1993; Robinson *et al.*, 1993), although other nomenclatures have emerged. We revert to the standard cyprinid nomenclature to stress class homologies. The zebrafish retinal mosaic is patterned with unequal double cones: Red VP570 opsin in the long double (LD) and green VP480 opsin in the short double (SD) members. Zebrafishes do not possess long single (LS) cones homologous to those of the goldfish, carp, roach or rudd. The blue VP415 opsin cones of zebrafish are structurally and, as will be shown, metabolically homologous to the short single (SS) cones of other cyprinids. Further, the UV VP362 opsin cones of zebrafish are structurally and metabolically homologous to the miniature short single cones (MSS) of the goldfish.

Cyprinids display pure rod and pure cone horizontal cells (Stell, 1975; Stell & Lightfoot, 1975) and the latter possess large axon terminal extensions (see review by Djamgoz *et al.* (1995)). The red cone dominated, monophasic H1 horizontal cells have a distinctive $\gamma+$ signature; spectrally opponent H2 and H3 (collectively termed Hx) horizontal cells have a $\gamma-$ signature; rod horizontal cells have a $\gamma-$/E+ (glutamate-positive) signature (Marc *et al.*, 1995; Watt *et al.*, 2000).

Cyprinid bipolar cells have mixed rod-cone or pure cone inputs (Scholes & Morris, 1973; Famiglietti *et al.*, 1975; Scholes, 1975; Ishida *et al.*, 1980a; Sherry & Yazulla, 1992). ON-center mixed rod-cone bipolar cells form a superclass separable into three classes (Bb1, 2, 3) in sublamina b (the proximal half) of the inner plexiform layer. OFF-center mixed rod-cone bipolar cells form a superclass separable into two classes (Ba1, 2) that terminate in the distal half of the inner plexiform layer, sublamina a. Homologues of the Ba and Bb superclasses are clearly present in the zebrafish retina. The remaining large cohorts of pure cone Ca, Cb, and Cab bipolar cells are complex and more difficult to compare, but evidence of C bipolar cell superclasses is strong. No satisfactory terminology has emerged for the complex amacrine cell and ganglion cell populations. Finally, the inner plexiform layer is highly structured and we use the *level* terminology (Ammermuller & Weiler, 1981; Marc, 1986) to specify position: the amacrine cell layer proximal border is level 0, and the ganglion cell layer distal border is level 100, with all other loci linearly scaled between.

RETINAL PREPARATIONS

These methods follow those described previously (Marc *et al.*, 1995). Light-and dark-adapted laboratory raised adult unsexed *Danio rerio* were deeply anesthetized in MS-222 aquarium water and cervically transected. Both eyes were rapidly removed, the corneoscleral junction immediately pierced with a 27 ga needle tip and flushed with room temperature fixative: 2% glutaraldehyde, 1% paraformaldehyde, 0.08 M phosphate buffer, 3% sucrose, 1 mm $MgSO_4$, pH 7.4. The anterior segment was removed and the eyecups fixed overnight. Speed of dissection was important for retention of GABA in horizontal cells and glycine in amacrine cells. All tissue was processed for resin embedding as previously described (Marc *et al.*, 1990).

SPECIMEN PREPARATION AND IMMUNOCYTOCHEMICAL VISUALIZATION

Each sample was sectioned serially at 250 nm onto 12-spot Teflon-coated slides (Cel-Line, Fisher Scientific). The immunocytochemical and IgG production procedures were as described previously (Marc *et al.*, 1995). The samples were serially probed with IgGs targeting alanine, aspartate, glutamate, glutathione, glutamine, glycine, GABA, and taurine obtained from Signature Immunologics Inc[1] (Salt Lake City, Utah). The selectivities of these probes have been extensively tested against their naturally occurring fee amino acid targets, as well as against a spectrum of other amines (Marc *et al.*, 1995; Marc & Jones, 2002). In practice, their selectivities for their target haptens is >3 log units in most cases. The relevant exceptions are as follows. (1) Anti-L-glutamate is only 8-fold selective for L-glu versus the metabolite N-acetyl-aspartyl-glutamate. However, N-acetly-aspartyl-glutamate is N-blocked and not captured by aldehyde fixation. (2) Anti-GABA is only 80-fold selective for GABA versus β-alanine. This poses no significant cross-contamination, as discussed in Marc and Jones (2002)). (3) Anti-glutathione does not discriminate between reduced and oxidized forms; it reports only total glutathione content. Primary IgG signals were detected with goat-anti-rabbit IgGs adsorbed to 1nm gold particles (Pharmacia-Amersham Intl plc) and visualized with silver intensification using the silver-intensification protocol of Kalloniatis and Fletcher (1993). The silvering process was run at 30°C for 240 sec before quenching. All preparations received identical probing and visualization treatments, yielding 2 orders of magnitude of detection range with differential concentration sensitivity as low as 40 μM based on artificial standards. Standard and augmented single letter amino acid codes are sometimes used to denote alanine (A), aspartate (D), glutamate (E), glutathione (J), glycine (G), glutamine (Q), and taurine (τ) and GABA (γ).

IMAGE CAPTURE

All images of immunoreactivity were captured as 8-bit frames under fixed CCD camera gain and gamma as previously described (Marc, 1999a). The objective lenses were either 20× apochromats or 40× oil planapochromats, yielding resolutions ranging from 485 nm/pixel to an over-sampled maximum of 126 nm/pixel; most images were captured at 243 nm/pixel, as this provides the maximum practical resolution and optimal pixel density for optical microscopy. Silver-intensified immunogold labelling produces *density-scaled* images, and linear image inversion produces *intensity-scaled* images that are linear with $\log_{10}$(concentration) of the target molecule over a range of 0.05–10 mM.

IMAGE ARCHIVE CREATION

The central step in molecular phenotyping is the creation of permanent data archives from registered image sets. Registration is a highly demanding operation and cannot be achieved by simply shifting images manually. Computational classification demands optimized data quality, which is a function of alignment precision. Even simple linear alignment at unit scaling requires both translations and rotations calculated from good control points. These are often difficult to identify in images that have complex or weak correlations in signature space. If there is any small shear due to section stretch, it cannot be managed by manual adjustment. Large sections often possess visually undetectable distortions that can corrupt linear alignments. A section 1 mm in extent with a mechanical stretch distortion of only 1%, embeds a propagating misalignment reaching 10 μm (two bipolar cell widths) in a linearly scaled image. This means that, starting from one "good" region of alignment, the data will be misaligned by a full cell halfway across the image and, in practice, over 90% of the image data will be useless. As we require <200 nm (<1 pixel) alignment precision, nonlinear spatial transform algorithms are essential. For similar reasons, automatic image alignment algorithms have never been successfully implemented. The most effective tool, in our hands, has long been remote sensing registration code from PCI Geomatica® (Richmond, Hill, Ontario, Canada).

Registered image archives were generated in two steps: registration of arrays of image tiles into mosaic channels (Fig. 1a) and registration between mosaic channels (Fig. 1b). Most of our data images subtend 187 μm × 140 μm, so a patch of tissue scaled at >40 tiles/mm^2 assembles into a single channel image of >17 megabytes (Mb)/mm^2. Thus 10 registered mosaics form a single ≈170 Mb/mm^2 file of aligned image channels at optimal resolution. Alignment was initiated by manual selection of small set of corresponding control points in master and slave images. These need not be at final precision, as the registration algorithms predict the best next slave image coordinate point based on Nth-order geometric models or local spline methods. Even with such predictive algorithms, it is extremely difficult to register images that lack strong signal differentiation or clear landmarks and the creation of datasets can take many weeks of work. Registered files are then exported from the proprietary file structure of the PCI Geomatica® code into a standard TIFF 6.0 format as a list of alpha channels.

Image classification based on small molecules has been detailed previously but follows this general scheme: (1) Isodata clustering of N registered channels, (2) theme map generation, (3) histogram exploration and, if necessary, (4) deconvolution and theme map correction. After theme map generation, univariate and bivariate signal histograms were explored for each emergent class. In cases where a signal was clearly multimodal, the histogram was deconvolved and each cell recoded. Classes resolved by clustering were formally *statistically separable* and those resolved by deconvolution were formally *statistically significant* (see below). Classifications were performed using the isodata algorithm (PCI Geomatica®) and data explored with applications written in IDL® (Research Systems Inc., Boulder, CO). An overview of and reference lists for pattern recognition methods are provided in Marc *et al.* (1995). The isodata algorithm resembles simpler K-means clustering methods but adds heuristic splitting of high variance classes and merging of highly overlapping classes (Ball & Hall, 1967). When the means and covariances for a set of N-dimensional data allow classifying a sample of those data into distinct classes, we call them *statistically separable*. The probability of error (p_e), estimated from the transformed divergences of the classes assuming equal *a priori* probability densities, is ≤ 0.01 for classes deemed *separable*. Separable classes are also inherently *statistically significant* classes. Natural classes need not be inherently separable by clustering and could have signature overlaps greater than required for separability yet still be statistically significant classes. Exploration of datasets with histogramming tools and deconvolution methods increases the sensitivity of analysis and permits detection of classes hidden in the complex array of N-dimensional data. Strongly separable cells can be removed first, simplifying analysis of the remaining data.

No set of monochrome images can visualize molecular mixtures or partial signatures. Selected 3-dimensional class signatures were visualized as follows: (1) *rgb* maps encoding molecular signal triplets as additive red, green and blue signals (see appendix in Marc *et al.* (1995)); (2) **cmy** maps encoding signals as subtractive cyan, magenta, and yellow triplets; and (3) *rgb*/**cmy** mixtures when visualization of classes is facilitated by a complex color space (*e.g.*, *r*-**m**-*b*, additive red, subtractive magenta, additive blue). The discriminations made by triplet mapping can be powerful, as quantitative colocalizations can be displayed, but no single triplet can capture the characteristic signatures of all cells. Sets of triplets are required and the particular triplets must be chosen carefully. After all, there are 56 unique triplets possible in 8-space, and four possible display conditions (pure *rgb*, pure **cmy**, and two mixtures) yielding over 200 "looks" at the same data, ignoring color permutations. Visually searching all these displays for correlations is ineffective and non-quantitative, but the right ones can reveal populations and relationships visible no other way.

Classification, as described above, is one proper solution to acquiring real numbers that characterize cell classes. *Classifications* were visualized as formal theme maps, where each color or other code represents a statistically significant group (class or superclass). The actual N-dimensional signatures were displayed as stacked bivariate *2N-plots* (Marc & Jones, 2002). Classical single bivariate plots display the area occupied by a class in a two dimensional chemical space. 2N-plots superimpose bivariate plots, capturing pairwise most of the features discriminating cell groupings in an N-space (Fig. 2). This method was inspired by the parallel coordinate space described by Inselberg and Dimsdale (1990). Pairs of signals were displayed as class means bounded by 1 SD margins. The XY axes spanned 0.1–10 mM with logarithmic scaling and the XY pairs were color-coded: grey AJ, gold Eγ, cyan DQ, and magenta Gτ. The class means were acquired after first transforming the raster-based image sets to *object* sets of

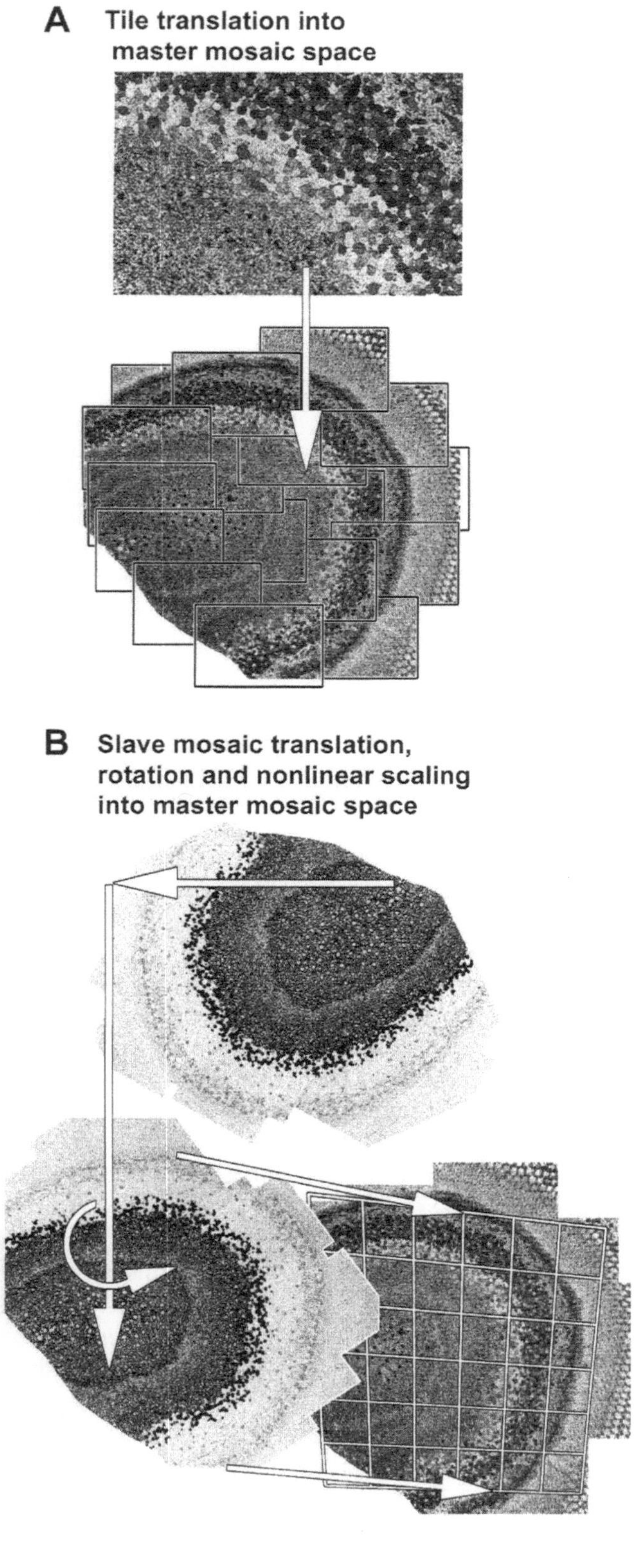

Fig. 1. Assembly of mosaics and registered channel sets. A. Single image tiles are spliced together with <200 nm error to form a large mosaic at optimal light microscope resolution (<250 nm). Each mosaic ranges from 5–20 Mb in size. This example illustrates the embedment of a single glutamate signal tile into an oblique section glutamate signal mosaic. B. Each mosaic in a series is precision-registered with all other mosaics, including corrections for shear and local warping. This example illustrates the translation, rotation and nonlinear scaling of a slave GABA signal mosaic to register it with a master glutamate signal mosaic. The grid represents the kind of distortion encountered with some large sections.

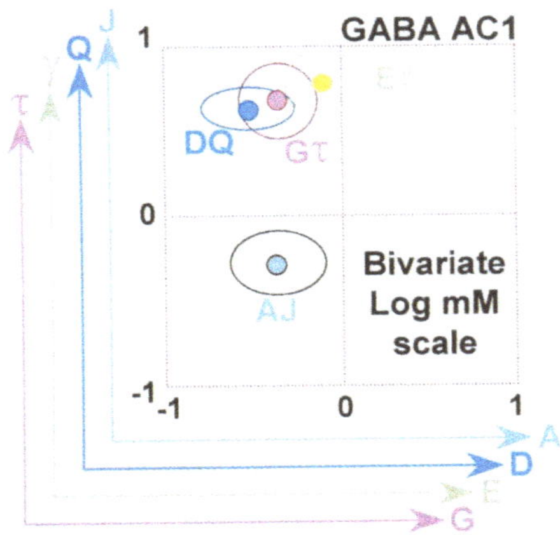

Fig. 2. N-plot for the ADEGJQ$\tau\gamma$ basis set. This N-plot was derived from the molecular phenotype data of class 1 GABA amacrine cells (GABA AC1). Each dot encodes a bivariate mean on a logarithmic mM scale (*i.e.* $-1 = 0.1$ mM, $0 = 1$ mM, $1 = 10$ mM), and is bounded by a 1 standard deviation ellipse. Four pairs of points represent a projection of 8-space into 2-space, and distinctions among classes are thus represented by the differing patterns of signals in N-plots (see Marc & Jones (2001)). The XY pairs are arbitrarily formed but constant throughout this manuscript: AJ (grey), DQ (cyan), Eγ (gold), Gτ (magenta). In this plot, it can be quickly determined that the mean A, D, G, E and J values are less than 1 mM in this class, because their coordinates are all below the log 0 mM line. Conversely, the mean Q, γ and τ values are all $\gg$1 mM.

classified cells and the means and variances calculated from cell numbers rather than pixel numbers.

Average cell sizes and cell spacings were acquired with the object utilities in ImagePro (Media Cybernetics, Silver Spring, MD). All cell sizes in each sample were measured only from individual 250 nm sections, such as the glutamate channel, so cell size estimates contain no registration errors and require no use of dissector or other stereological correction methods. Some classified cells often appeared to form regular arrays, and these were numerically characterized by their *conformity ratios* (also known as regularity indices): the ratio of the class mean nearest-neighbor distance to its standard deviation. The significance of the deviation of a conformity ratio from that predicted for a random pattern was determined from significance charts (Cook, 1996).

FIGURE PREPARATION

All images are digital, assembled from the raw data captured by CCD camera (see *Image Analysis* above). Selected frames of raw Tagged Image Format (*.tif) files were extracted for display and contrasts were adjusted with linear remapping to optimize fidelity of prints to the original microscopic images. All final images were prepared in Adobe PhotoShop® 6.01.

Results

The general metabolic phenotypes of various zebrafish retinal cells are similar to most vertebrates, including mammals, but the specific classes that they fall into of define zebrafish as cyprinid fishes as surely as any taxonomic feature: a row patterned cone mosaic, at least three and likely four horizontal cell classes, bipolar cell classes with characteristic Ba and Bb terminal morphologies, a ≈2:1 γ+/G+ amacrine cell population ratio, the presence of glycinergic interplexiform cells, and simple ganglion cell layer with minimal coupling signals.

THE GENERAL PLAN OF THE RETINA

Both cellular organization and metabolic differentiation are well illustrated by the patterning of glutamate signals in the dark-adapted zebrafish retina shown in Figure 3. This image is essentially a quantitative map of glutamate distribution and clearly distinguishes the cell types of the retina. The adaptation state is an important topic to which we will return. The highest neuronal glutamate levels are expressed by some bipolar cells, their dendrites in the outer plexiform layer and their axon terminals in the inner plexiform layer. The signal strength corresponds to 5–10 mM free glutamate in the cytosol. This high concentration is not unique to neurons, however, as the endothelial cells forming the radially arrayed vitreal blood vessels that line the proximal margin of the retina contain as much or more glutamate. The remaining bipolar cells, most ganglion cells and dark-adapted long (LD) and short (SD) double cones and rods express roughly 2–5 mM glutamate. The next lower range (0.5–1 mM) includes the short (SS) and miniature (MSS) short single cones, the RPE, and the horizontal cells. The lowest retinal glutamate levels (<0.3 mM) are displayed by amacrine and Müller cells.

Two important concepts emerge from the image in Figure 3. First, the cells of the zebrafish retina are exceptionally well laminated in comparison to most other vertebrates, including its cyprinid relatives. The somas of the bipolar cells are displaced some 5 μm from the outer plexiform layer by the interposed somas of the horizontal cells and 10 μm from the inner plexiform layer by the interposed amacrine cells; thus distinct horizontal, bipolar and amacrine cell layers comprise the inner nuclear layer proper. As will be shown, the somas of the Müller cells are less precisely positioned, but preferentially occupy the border between the bipolar and amacrine cell layers. The inner plexiform layer also displays a precise division into sublayers occupied by characteristic cyprinid bipolar cell cohorts. Classically, sublaminas *a* and *b* represent zones dominated by OFF-center and ON-center rod and long-wave cone processing, respectively. This is reflected in the laminar partitioning of characteristic size groupings of bipolar cell terminals: (1) the large 5–7 μm diameter terminals of ON-center Bb bipolar cells at level 80–100; (2) the medium 3–5 μm terminals of OFF-center Ba bipolar cells, which are further divisible in to classes Ba1

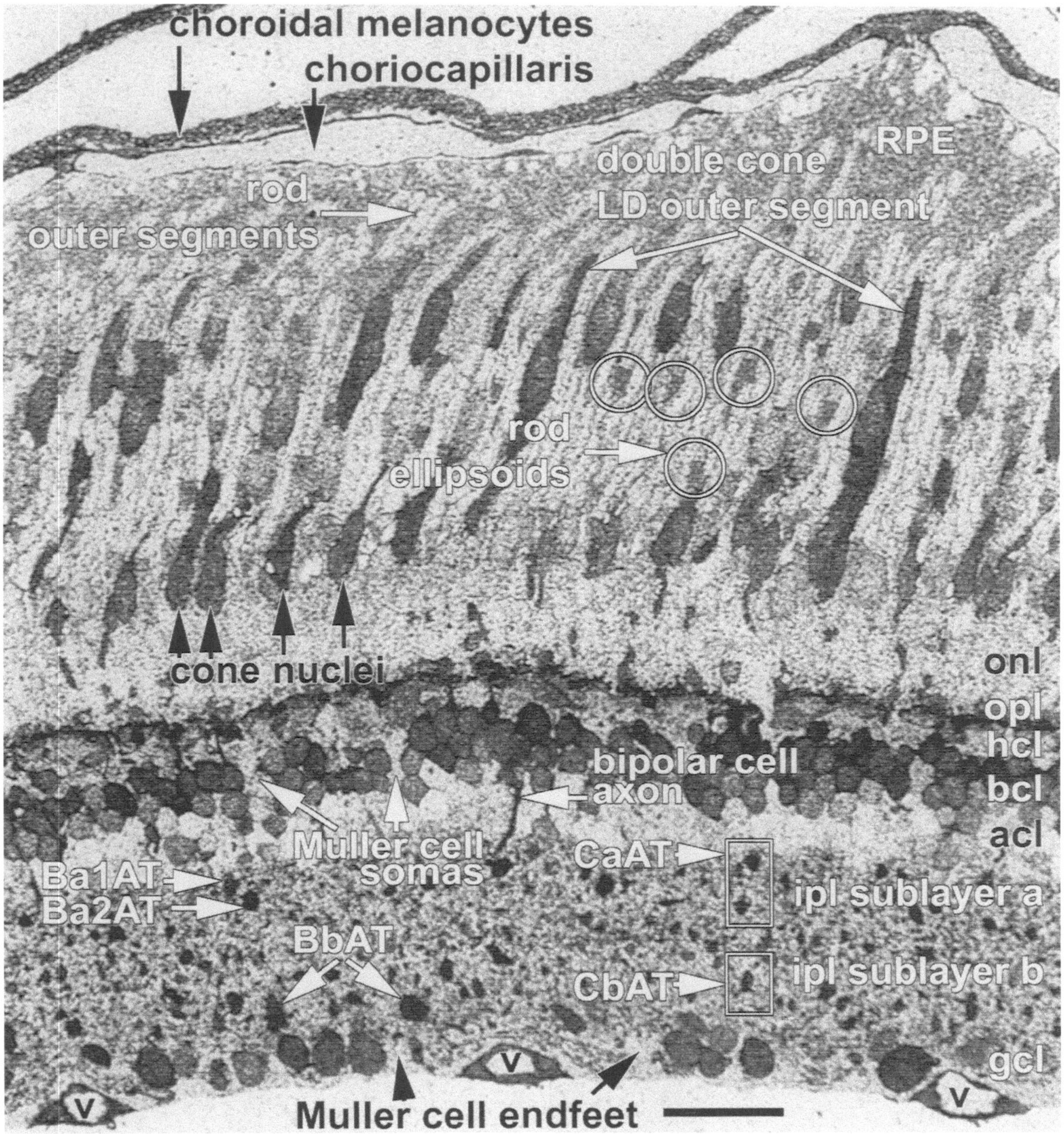

Fig. 3. Glutamate signals in the dark-adapted zebrafish retina. This single image displays virtually all choroidal, RPE and retinal cell types by virtue of their varying contents of free glutamate. Rod ellipsoids are circled and key structures labeled. Section thickness: 250 nm. Scale 20 μm. Abbreviations: acl, amacrine cell layer; Ba1AT, class Ba1 bipolar cell axon terminal; Ba2AT, class Ba2 bipolar cell axon terminal; BbAT, class Bb bipolar cell axon terminal; CaAT, class Ca bipolar cell axon terminal; CbAT, class Cb bipolar cell axon terminal; bcl, bipolar cell layer; gcl, ganglion cell layer; hcl, horizontal cell layer; ipl, inner plexiform layer; LD, long double cone member; onl, outer nuclear layer; opl, outer plexiform layer; RPE, retinal pigmented epithelium; v, radial vitreal blood vessel.

and Ba2; and (3) profuse examples of small pure cone bipolar cell varicosities (Ca, Cb). The separations and precisions of the sublayer borders are actually superior to that of the best-studied cyprinid, the goldfish. Even though the adult zebrafish eye is some 6-fold smaller in area than a 3–4 year old goldfish, the inner plexiform layer is still as thick as the central goldfish retina inner plexiform layer, $\approx$30–35 μm. This suggests that the zebrafish contains as many bipolar/amacrine/ganglion cell types *per mosaic unit cell* as the goldfish, despite its small size.

The second concept emerging from Figure 3 is that each retinal cell type shows a distinctive, stable, but *dynamic* metabolic profile: the profile is the consequence

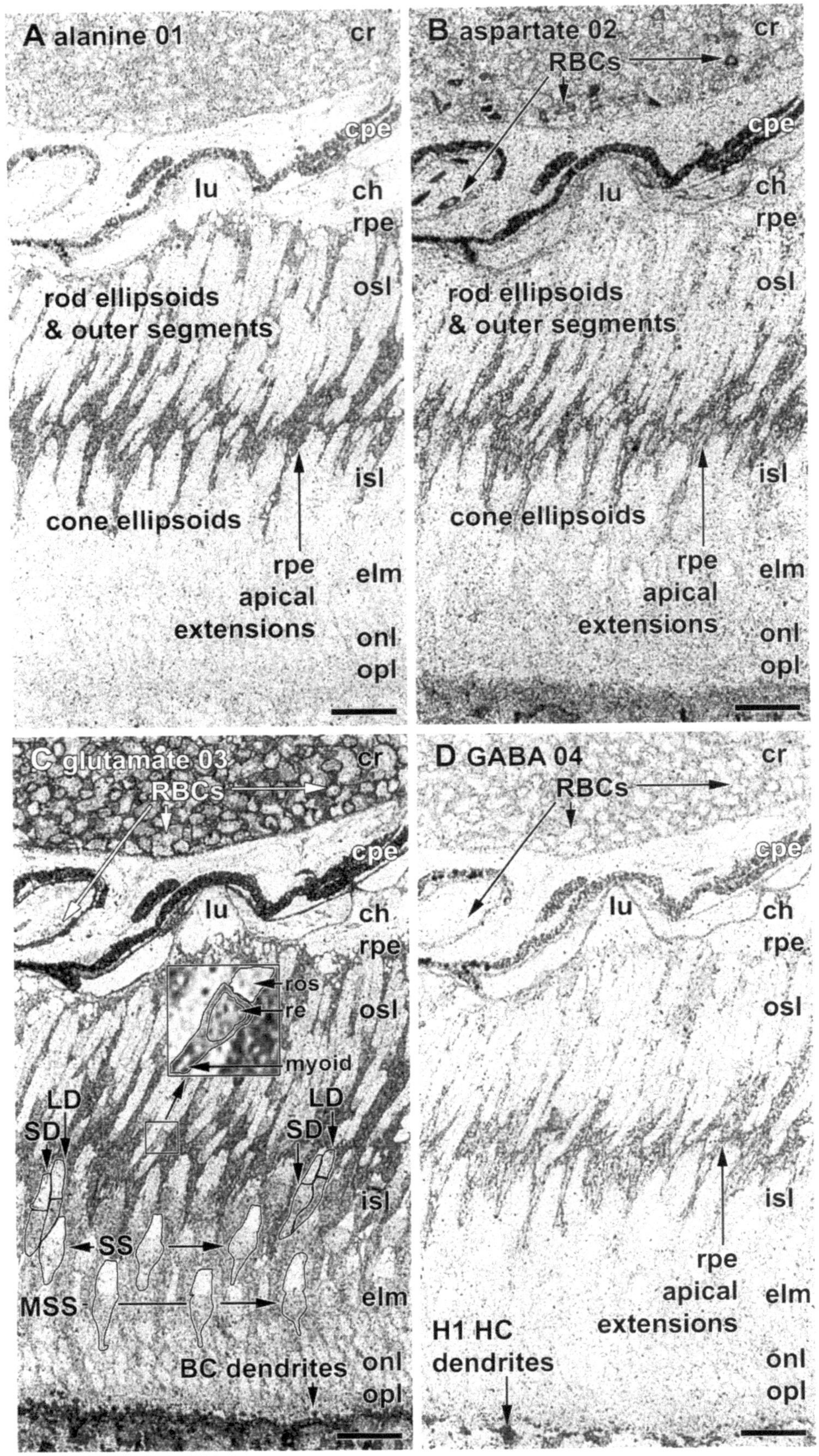
A alanine 01
cr
cpe
lu
ch
rpe
osl
rod ellipsoids
& outer segments
isl
cone ellipsoids
rpe
apical
extensions
elm
onl
opl
B aspartate 02
RBCs
C glutamate 03
ros
re
myoid
LD
SD
SS
MSS
BC dendrites
D GABA 04
H1 HC
dendrites

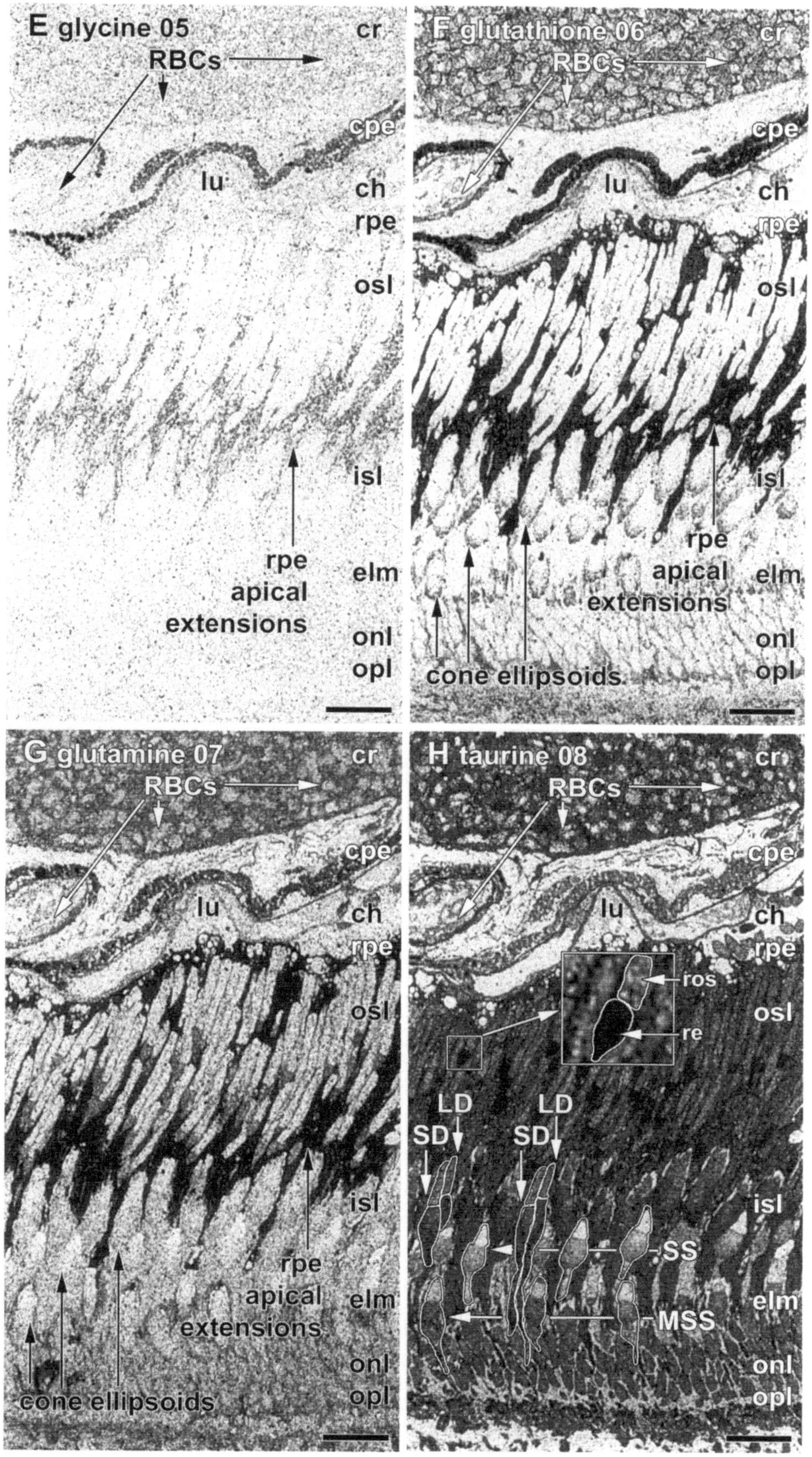
E glycine 05
RBCs
cr
cpe
lu
ch
rpe
osl
isl
rpe apical extensions
elm
onl
opl
F glutathione 06
RBCs
cr
cpe
lu
ch
rpe
osl
isl
rpe apical extensions
elm
onl
opl
cone ellipsoids
G glutamine 07
RBCs
cr
cpe
lu
ch
rpe
osl
isl
rpe apical extensions
elm
onl
opl
cone ellipsoids
H taurine 08
RBCs
cr
cpe
lu
ch
rpe
ros
re
osl
LD
SD
isl
SS
MSS
elm
onl
opl

of multiple fluxes. For example, it is well-known that the most potent glutamate transport in the retina is displayed by the Müller cells, both due to their high expression of transporters and their massive surface area, which includes both the proximal and distal diffusive seals of the neural retina. Even so, the Müller cells display the lowest steady-state glutamate content, due to their expression of very high levels of glutamine synthetase. Either chemical inhibition of glutamine synthetase (Pow & Robinson, 1994) or pathologic loss of glutamine synthetase expression (Marc *et al.*, 1998a) evokes a rise in intracellular glutamate in Müller cells that reaches the levels of glutamatergic neurons. As we will see, these and other cells exhibit either static or dynamic metabolic phenotypes depending on the linkage of source and sink processes to adaptational or respiratory states. We will now explore the molecular phenotypes of retinal cells in a distal → proximal fashion, using molecular phenotypes (where necessary) to distinguish cell classes.

THE LIGHT-ADAPTED OUTER RETINA

The outer retina, in a metabolic sense, is a compartment sealed between the basolateral tight junctions of the RPE and the microvillar tight junctions of the Müller cells (Fig. 4). The RPE serves as a distal transport monolayer between the extracellular plasma in the lumens of the fenestrated choriocapillaris endothelia and the extracellular subretinal space. The zebrafish, possesses a teleost apomorphic organ, the *choroidal rete*, which mediates counter-current amplification of oxygen tension. By analogy with larger fishes, this structure is expected to generate pO_2 values of >300 mm Hg, 3 to 4 (Fonner *et al.*, 1973) times the saturated arterial values of mammals. The zebrafish retina likely has the same ultrahigh oxygen demands as other teleosts and is thus not a plausible general model for oxygenation in the outer retina of mammalians. The diverse cellular composition of this compartment displays subtly diverse molecular phenotypes when probed with the basis metabolite set of alanine, aspartate, glutamate, glycine, glutathione, glutamine, taurine and GABA (abbreviated ADEGJQ$\tau\gamma$).

Alanine levels are extremely low in the outer retina, reaching their maximum value in the RPE of 0.5–1 mM, and roughly half of that in the endothelia of the choroidal rete (Fig. 4A). To put this minor elevation in perspective, the photoreceptors, with perhaps the highest sustained rates of retinal protein synthesis, maintain less than 50 μM alanine. What, then, is the purpose of a metabolic pathway that maintains such an elevated level of alanine in the RPE? Alanine is not an essential dietary amino acid in fishes, but is a key metabolite in the maintenance of pyruvate pools via general transamination paths and is a source of amino groups (Michal, 1999). As will be considered below, one source of retinal glycine synthesis may involve transamination of alanine and glyoxylate.

Aspartate signals are a little lower than alanine signals in most cells of the outer retina, with the exception of nucleated red blood cells, evident both in the choriocapillaris and the choroidal rete (Fig. 4B). In contrast, *glutamate* signals are actually weaker than aspartate in the red blood cells. The vascular endothelia, the RPE and photoreceptors all express elevated glutamate levels, but with some significant differences (Fig. 4C). The highest glutamate levels in the light-adapted outer retina are maintained by the RPE at ≈1 mM, while all photoreceptors maintain lower levels, especially the SS cones. As in the goldfish (Marc *et al.*, 1995), light-adapted zebrafish photoreceptors maintain glutamate levels in the 100–800 μM range. We are accustomed to the idea that cells must maintain high cytoplasmic levels of their specific neurotransmitters, yet it is evident here that only very low intracellular levels of glutamate

Fig. 4. (Previous 2 pages) Single channel, registered ADEGJQ$\tau\gamma$ signals in the light-adapted photoreceptor layer. The index number following the signal label on each image indicates its position in the series. A, alanine 01. Only the RPE has significant signals. The signal strength in the choroidal pigmented epithelium reveals the basal optical density produced by endogenous melanin. B, aspartate 02. The RPE aspartate content is low, exposing the positioning of melanin in the RPE apical extensions. Conversely, the choroidal pigmented epithelium contains high levels of aspartate, as do RBCs. C, glutamate 03. Glutamate signals are elevated in the RPE, choroidal pigmented epithelium and all choroidal endothelia. Photoreceptor levels are modest and slightly less than the RPE. The two inset boxes are 10 μm square and show the details of glutamate distribution in a rod. D, GABA 04. Only basal signals are present, but for the dendrites of horizontal cells. E, glycine 05. Only low levels of glycine are present. F, glutathione 06; The RPE and choroidal pigmented epithelium display extremely high glutathione content, with lower more complex patterns in other cells. G, glutamine 07. Similar to glutathione, glutamine reaches extremely high concentrations in the RPE and choroidal pigmented epithelium. H, taurine 08. Taurine is the most abundant amino acid in the outer retina and is enriched in all cell types but for the choroidal pigmented epithelium and short single cones. The two inset boxes are 10 μm square and show the details of taurine distribution in a rod. Section thickness: 250 nm. Scale 20 μm. Abbreviations: BC, bipolar cell; ch, choroid; cpe, choroidal pigmented epithelium; cr, choroidal rete; elm, external limiting membrane; H1 HC, class H1 horizontal cells; lu, vascular lumens in the choriocapillaris; isl, inner segment layer; MSS, miniature short single cones; osl, outer segment layer; re, rod ellipsoid; ros, rod outer segment; SD, short double cone members; SS, short single cones; RBCs, nucleated erythrocytes; other abbreviations as in Figure 3.

are needed to support photopic glutamatergic synaptic transmission in the outer plexiform layer. This implies that the efficacy of vesicle loading by glutamate is very high.

The non-protein amino acid *GABA* is best known as a vertebrate inhibitory neurotransmitter and, consistent with this, its signals are extremely low in the outer retina (Fig. 4D). The elevated GABA signal in H1 horizontal cell dendrites gives visual scaling. There is $\approx 100\ \mu M$ GABA in the endothelia of the rete and the RPE, though these cells do not express detectable levels of glutamic acid decarboxylase (GAD) by qualitative immunocytochemical criteria. But it is critical to recall that GABA is a primary respiratory metabolite in plants and bacteria and that GAD expression directly couples to the maintained level, so low levels of GABA are not necessarily surprising in non-neural cells. Furthermore, GABA is an obligate intermediate in the polyamine degradation pathway (Michal, 1999), and cells with high polyamine synthesis rates might accumulate low levels of GABA. *Glycine* is also best known as an inhibitory neurotransmitter, but is clearly maintained at protein synthesis levels (50–300 μM) in the outer retina (Fig. 4E).

One of the most dramatic signal patterns in the outer retina is the partitioning of *glutathione* (Fig. 4F), which reaches 5–10 mM levels in the somas and apical extensions of the RPE. Glutathione levels are immeasurably low in photoreceptor outer segments, but reach about 0.5 mM in granular zones surrounding the proximal margin of the ellipsoid in all cone types. Moderate levels are expressed in the endothelium of the rete and higher levels in the choroidal pigmented epithelial stroma. The demand for glutathione can arise from three primary processes: (1) the need for thiol reducing equivalents in various metabolic reactions; (2) the need to quench reactive oxygen species formation; (3) the need for glutathione-coupled amino acid transmembrane transport. At the least, the RPE is likely to require the latter two functions to accompany its roles in degrading photoreceptor outer segment membranes rich in oxidizable fatty acids, operating in excessive oxygen tensions and mediating transcellular amine transport. The apical extensions of the RPE reach within 20 μm of the external limiting membrane. Müller cells also contain ≈ 1 mM glutathione, as evidenced by the outlining of the rod nuclei and a distinct band of glutathione-rich punctae at the level of the outer limiting membrane. Thus the subretinal space is bordered by cells that maintain exceptionally high glutathione levels.

The content of the amino acid *glutamine* somewhat resembles that of glutathione, though it is more abundant in rod inner segments than cone inner segments. The RPE contains over 10 mM glutamine: a level rivalled only by ganglion cells. In neurons, glutamine is thought to be a primary source of glutamate via phosphate-activated glutaminase in the inner mitochondrial membrane. The RPE clearly does not need to maintain this pool of glutamine just for glutamate synthesis and it is not clear how the RPE synthesizes or acquires it. Glutamine is, however, one of the major sources of amino equivalents for the generation of amino sugars and pyrimidine interconversions.

Finally, the non-protein acidic amine taurine is widely distributed in the outer retina and every cell contains supramillimolar levels. The highest levels are located in the ellipsoids of rod cells, followed by the RPE, the endothelia of the rete, and the LD and SD cones. The lowest taurine levels are maintained by the SS cones, as in other cyprinid fishes. These images also reveal that the outer segments of photoreceptors are possibly well-equilibrated with the cytosol of the inner segments. The taurine level is always lower in the outer segments but matches that in the inner segments: outer segments of cones with higher inner segment taurine are also taurine-enriched. Outer segments are composed of extremely dense arrays of rod discs or cone lamellae, with little intracellular space. Based on electron microscopic images, the cytosol of rod and cone outer segments likely comprises less than 1/3 of the outer segment volume, and apparent taurine signals should be correspondingly reduced even if fully equilibrated with the inner segment due to optical blurring. Taurine has never been found to have a significant metabolic or neurochemical role and is widely thought to be one of several small organic osmolytes involved in volume regulation. In addition to ion channel permeations that may change water distribution, the photoreceptors and the RPE of many aquatic vertebrates undergo large shape changes in light and dark adaptation that may require robust pressure regulation.

Though the outer retina is a relatively simple compartment, it is heterocellular, with 9 resident cell classes: four photoreceptor classes, two endothelial classes, two epithelial classes, and one glial cell class. Qualitative differentiation of these classes is gained by examining color triplet images. Mapping the triplet taurine $\rightarrow$ red, glutamine $\rightarrow$ green, glutamate $\rightarrow$ blue (abbreviated τQE $\rightarrow$ *rgb*), clearly distinguishes most of the cellular elements in a single step, with the yellow (τQ) signal of RPE cells setting them apart from the dark-to-bright red τ signal of photoreceptors and the bright green threads of the Müller cells in the outer nuclear layer (Fig. 5A). The endothelia of the choroids are a pale pink signal, while blood elements are only poorly visible. Changing the triplet mixture to DJE $\rightarrow$ *rgb* (Fig. 5B) reveals the high aspartate levels in red blood cells and contrasts the RPE against the darkened photoreceptors. Only glutathione-rich necklaces proximal to the cone ellipsoids identify them. The most complete separation of the RPE signature from all others cells is achieved with a τJQ $\rightarrow$ *rgb* mapping (Fig. 5C), which renders it virtually white by summing the high signals of each metabolite. The triplet values indicate that the free amine content of the RPE is around 25 mM, considering these three

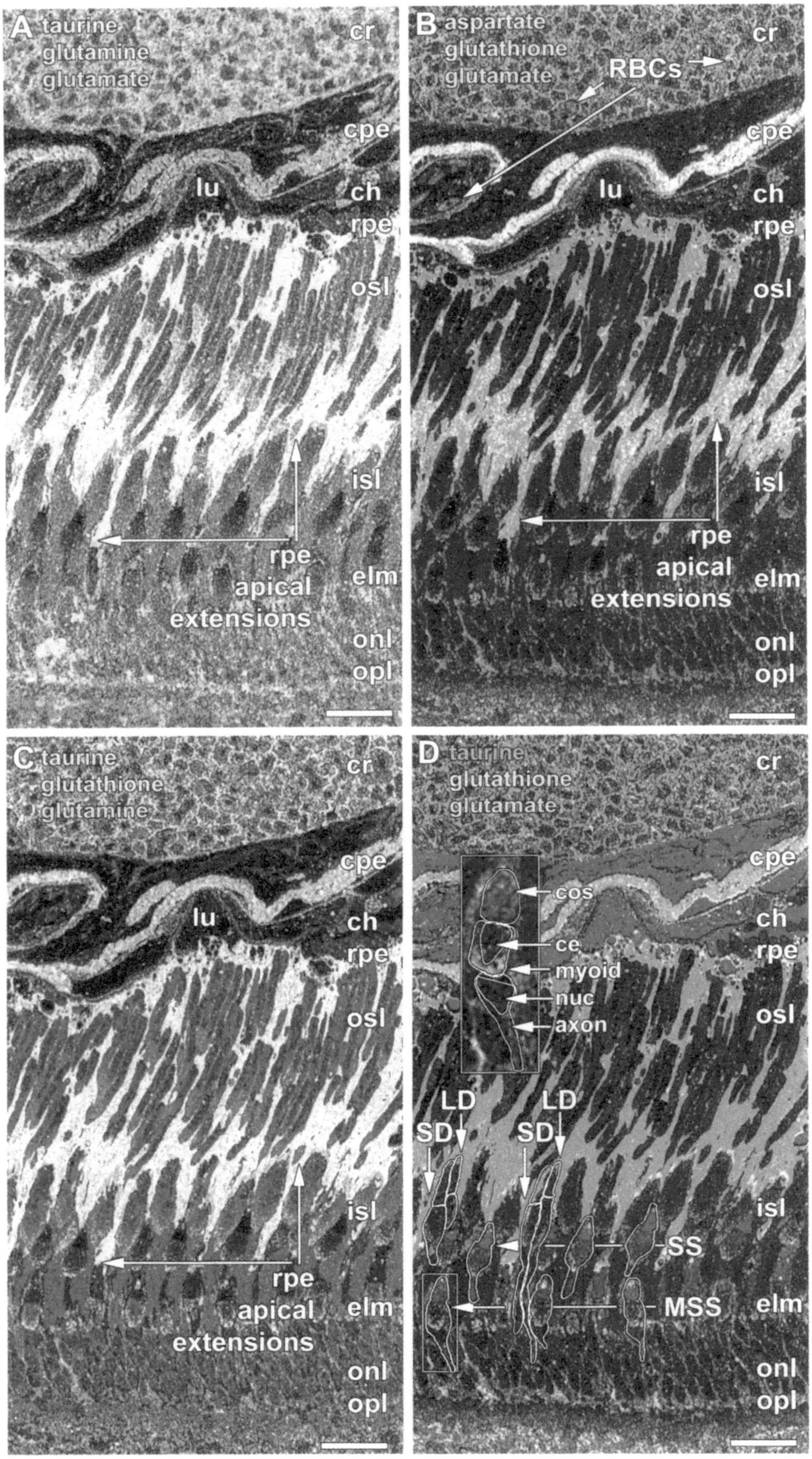
A taurine
glutamine
glutamate
cr
cpe
lu
ch
rpe
osl
isl
rpe
apical
extensions
elm
onl
opl
B aspartate
glutathione
glutamate
RBCs
cr
cpe
lu
ch
rpe
osl
isl
rpe
apical
extensions
elm
onl
opl
C taurine
glutathione
glutamine
cr
cpe
lu
ch
rpe
osl
isl
rpe
apical
extensions
elm
onl
opl
D taurine
glutathione
glutamate
cr
cos
ce
myoid
nuc
axon
cpe
ch
rpe
osl
LD
SD
LD
SD
isl
SS
MSS
elm
onl
opl

metabolites alone. We have measured most of the other free amino acids (*e.g.*, cysteine, lysine, leucine, arginine, serine, threonine, tryptophan and tyrosine) and found them to range from below 50 up to 250 μM. If all the remaining amino acids are of similar concentrations, they will total less than an additional 2–4 mM. Since the intracellular osmolarity of freshwater fish retinas is between 260–300 mOsm, one can reasonably assume that about 10% of that is derived from the total free amine pool. Free sugars, carboxylates, nucleotides, etc., likely represent at least another 10%, but the exact measurements remain to be made.

Finally, remapping into a complex space by τJQ $\rightarrow$ **c***gb* (subtractive cyan, additive green and blue) strongly delineates the metabolic compartments surrounding and within the cones, as illustrated by the inset in Figure 5D. A MSS cone has distinct metabolic mixtures in its outer segment, ellipsoid, myoid and axonal regions. Part of this is due to the specialized accumulation of glutathione (probably in vesicular compartments), the accumulation of taurine in the cytosol and mitochondria and the small cytosolic volume of the outer segment.

THE DARK-ADAPTED OUTER RETINA

One might consider this depth of metabolic profiling an academic exercise were it not for the facts (1) that there are no proven norms for neuronal metabolism *at the cellular level*... we are discovering them... and (2) that some signatures change upon adaptation, indicating rapid coupling between the cellular and system states. By revisiting the zebrafish retina following dark adaptation, we discovered that the glutamate contents of rods and some cones increase several-fold (Fig. 6A). The glutamate content of the RPE, however, remains unchanged. Conversely, the glutamine levels of the photoreceptors drop (Fig. 6B).

A comprehensive summary of these changes, based on monochrome and color triplet images, would be qualitative and tedious. But we can summarize all the quantitative data for various cell types by exploiting N-plots as described in the Methods. Figure 7 displays N-plots for the RPE, rods, MSS, SS, SD, and LD cones. All 8 basis molecules are plotted in 4 superimposed XY pairs (AJ, DQ, Eγ, Gτ) in both light and dark-adapted states. MSS cones change little, with a 2-fold decrease in glutamine content. The shift in SS cones is a bit more dramatic, with increases in glutamate and decreases in both glutamine and taurine levels. Rods show a large glutamate increase and a glutamine decrease but no evidence of taurine redistribution. The LD and SD cones show by far the most dramatic changes, with 6-fold increases in glutamate content up to 2–4 mM and concomitant drops in glutamine and taurine. The RPE serves as an internal control, with no change in glutamate content and minor (statistically insignificant) shifts in other signals. This is a more quantitative description of the process first described for the goldfish retina (Marc *et al.*, 1995). The meaning of these changes is unknown, but they

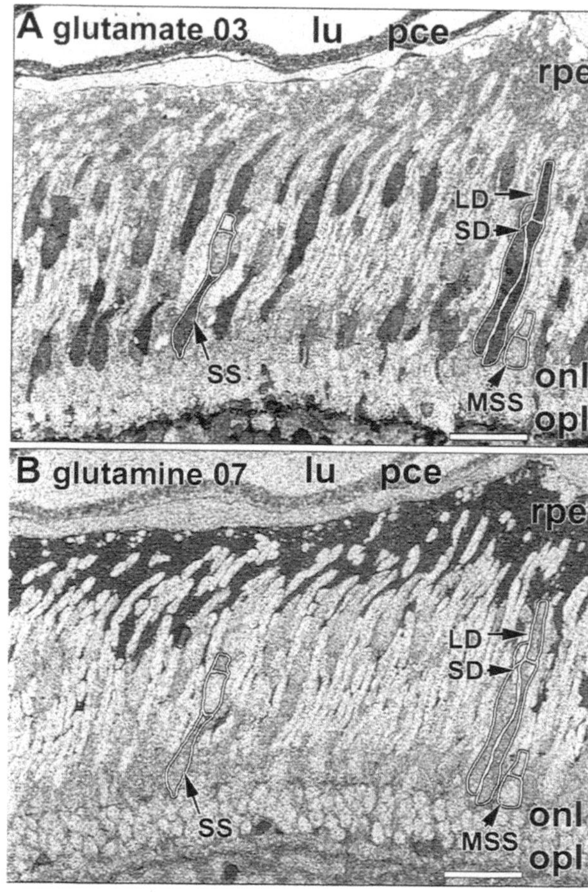

Fig. 6. Registered EQ signals in the dark-adapted photoreceptor layer. A, glutamate 03. The ellipsoids and myoids of LD and SD cones display large increases in glutamate content as they undergo photomechanical extension towards the RPE. B, glutamine 07. The same cones display decreases in glutamine content, while the RPE signals remain unchanged. Section thickness: 250 nm. Scale 20 μm.

Fig. 5. Triplet mapped basis set signals in the light-adapted photoreceptor layer. A, τQE $\rightarrow$ *rgb*. B, DJE $\rightarrow$ *rgb*. C, τJQ $\rightarrow$ *rgb*. D, τQE $\rightarrow$ **c***gb*, *i.e.* a complex mapping of subtractive cyan, additive green and blue. The two inset boxes are 10 μm $\times$ 30 μm and show the varied signal mixtures in MSS cone subcellular compartments. Scale 20 μm. Abbreviations: ce, cone ellipsoid; cos, cone outer segment; nuc, nucleus; other abbreviations as in Figures 3 and 4.

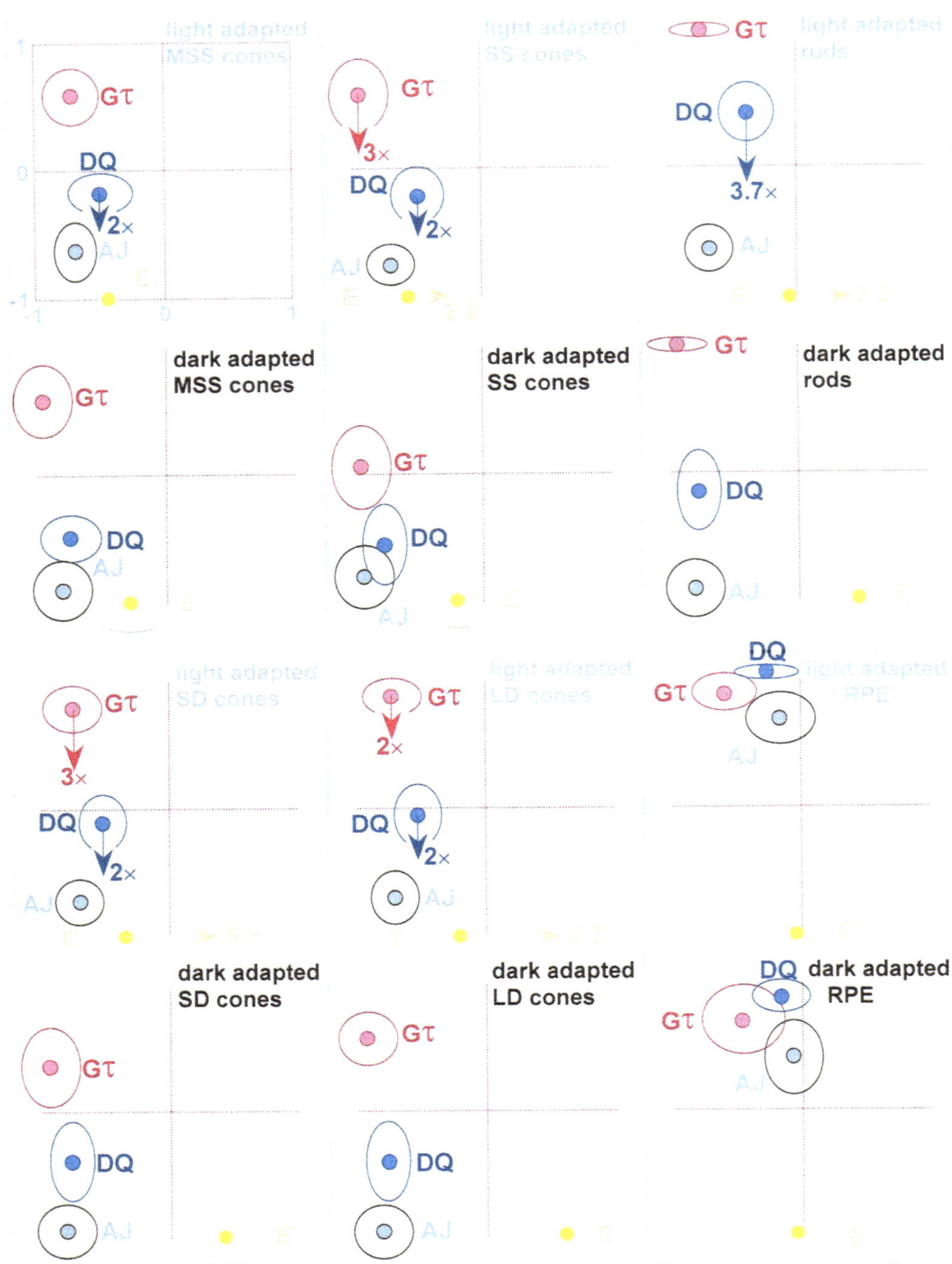

Fig. 7. Photoreceptor/RPE molecular phenotyping N-plots. All details as in Figure 2. Each cell class is plotted as a pair of N-plots, light-adapted above and dark-adapted below. The RPE shows no significant changes in signature upon adaptation, while rods, SD and LD cones show extensive increases in glutamate content and decreases in both glutamine and taurine levels.

parallel the dramatic adaptation-driven change in cell morphology displayed by LD and SD cones in teleost retinas. Dark-adaptation activates a tremendous increase in the surface area of the myoid and may involve volume changes large enough to demand compensatory responses. The smaller changes in SS and MSS cones are in accord with their smaller morphological alterations upon dark adaptation.

CONE ARRAYS

The strong signal differentiation of the dark-adapted cones also permits a unique chemical visualization of the cone mosaic in the zebrafish retina. Sectioned in dorsal retina at the level of the MSS cones, Figures 8 and 9 display the modest glutamate signal of MSS cones ellipsoids and the precisely arrayed high glutamate myoid LD-SD pairs next to one. Sandwiched between MSS

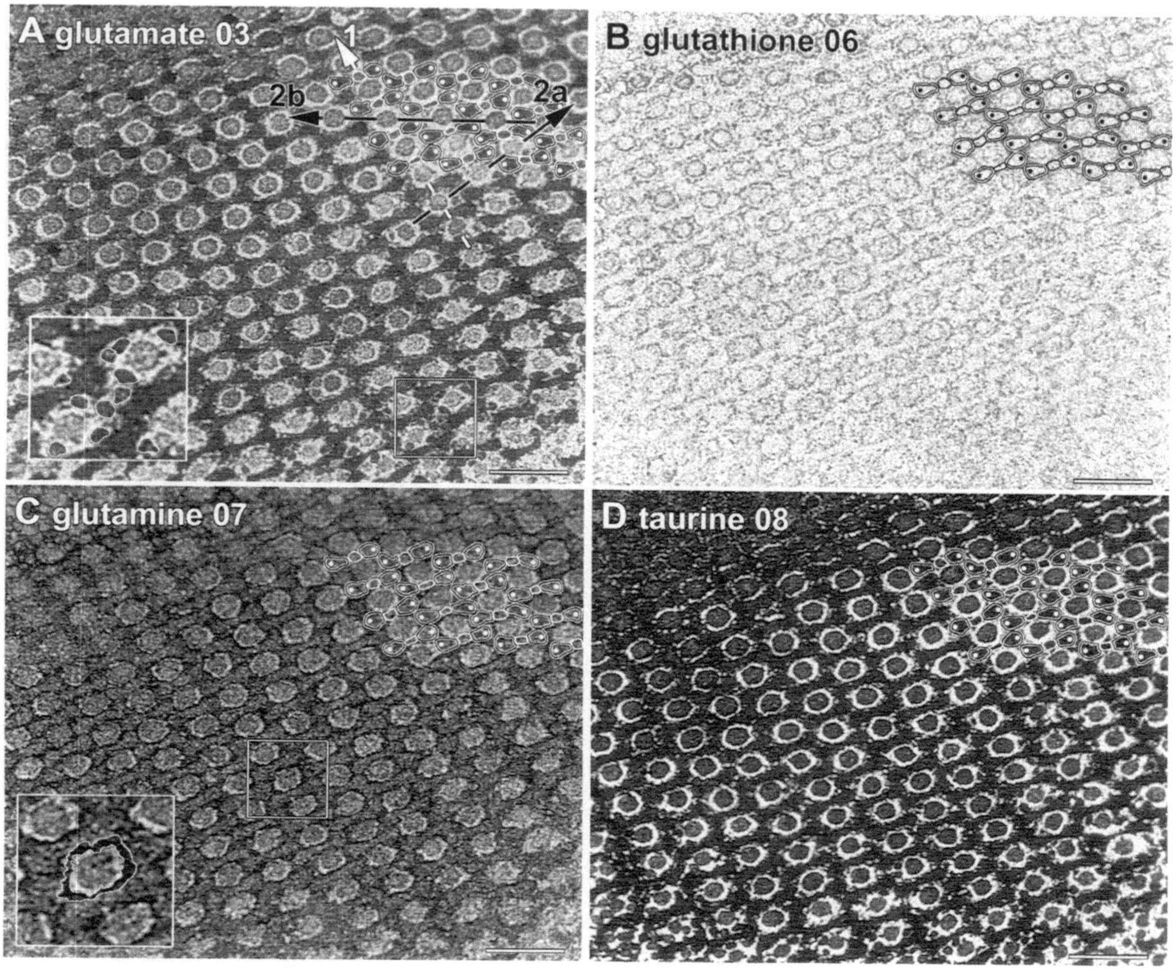

Fig. 8. Single channel, registered EJQτ signals in the dark-adapted cone mosaic. A, glutamate 03. The circular elements are the UV cones and surrounding extracellular matrix and the glutamate-rich bow-tie structures are the myoids of double cones (double outlined) aligned with the myoids of blue cones. According to the scheme of Raymond *et al.* (1993) and Robinson *et al.* (1993), the UV cones are always adjacent to the green SD cones, indicated by white dots. The primary (1) and secondary (2a,b) tiling axes are marked for a patch of the mosaic. The two inset boxes are 20 μm square and show the emergence of rod ellipsoids (outlined) in the distal portion of the dark-adapted cone mosaic. B, glutathione 06. The glutathione signals are precisely positioned within cone ellipsoids and the melanin-free RPE apicial extensions around all cone borders. C, glutamine 07. The highest contents are in the RPE apical extensions. The two inset boxes are 20 μm square and show the high glutamine signals of RPE processes outlining the MSS cone extracellular space. D, taurine 08. All cellular elements express high taurine levels, selectively exposing the extracellular space. Section thickness: 250 nm. Scale 20 μm.

cones are the small myoids of SS cones. Each double cone and SS cone myoid element is completely surrounded by a thin glutamine- and glutathione-enriched rind of RPE processes, while a 500–750 nm thick ring of extracellular matrix resides between the MSS cone ellipsoids and the RPE processes. The precision of the zebrafish cone mosaic appears to be superior to that of the goldfish. Both MSS and SS cones are present at density of 9400 cones/mm^2 (uncorrected for shrinkage) in this dorsal retinal sample, some 2–4× higher than the density of SS cones in the adult goldfish retina. Both cone placements have conformity ratios >13, far higher than the precision of cell patterning of any cell class in the mammalian retina. The double cone density = SS cone density × $\sqrt{2}$, just as in the goldfish retina. Though the mosaic appears as a pure row mosaic, the precise 1/2 cycle offset of neighboring rows forces the scaling to be the same as the pure square mosaic found in cyprinine and perciform retinas. The chromatic sampling originally proposed by Scholes (1975) holds as well for the zebrafish as for other cyprinids: when optimally focused on a long-wave target, lens chromatic aberration forces short-wave cone mosaics to be coarser. Any denser SS or MSS cone packing would be empty resolution.

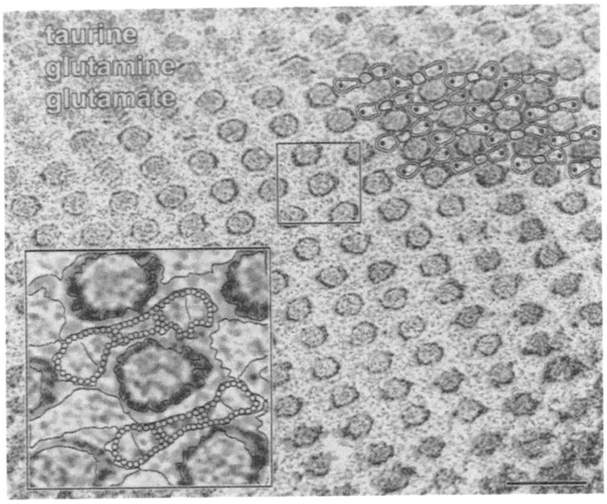

Fig. 9. Triplet mapped τQE → *rgb* signals in the dark-adapted cone mosaic. The separate zones of the mosaic tiling express different signatures. The two inset boxes are 20 μm square and show the continuum of green RPE processes (solid black outlines) around the pink SS and double cone myoids (blue, red and green dotted outlines), and the ≈1 μm dark ring of extracellular matrix around each MSS cone ellipsoid. Scale 20 μm. Tiling axes as in Figure 8.

THE INNER RETINA

The inner retina displays much greater signal differentiation than the outer retina due to its abundance of specialized neurons and their precise lamination. As with the outer retina, we can begin with a general survey of the molecular signals.

As in the outer retina, *alanine* signals are for the most part weak, but are clearly elevated in the inner plexiform layer and in a subset of cells in the amacrine cell layer (Fig. 10A). However, *aspartate* signals (Fig. 10B) are more differentiated in the inner retina, with a clear elevation in the horizontal cell layer and especially in rod horizontal cells (to be discussed below). Strong aspartate signals are also evident in small subsets of the bipolar and amacrine cell populations. Again, very distinctive partitioning of the retina is achieved with *glutamate* signals (Fig. 10C) with nearly all glutamate-rich bipolar cells perfectly sandwiched between the horizontal and amacrine cell layers, their terminals dispersed in patterns through the inner plexiform layer. The ganglion cell layer displays a high and fairly uniform glutamate signal, but gaps appear which will be later shown to be the somas of displaced γ+ amacrine cells. As in the outer retina, *glycine* levels are low in the inner retina, but the ≈0.7–2 mM signals of G+ amacrine cells sets them apart from all other cell types (Fig. 10D). There is a distinctive G+ band at level 50 of the inner plexiform layer, with diffuse punctae throughout the inner plexiform layer. No G+ somas were detected in the ganglion cell layer. *Glutathione* signals set the Müller cells apart from all neurons and reveal that Müller cell somas are predominantly positioned in the distal portion of the amacrine cell layer, next to the bipolar cell layer (Fig. 10E). The end-feet of the Müller cells form a J+ border at the vitreal margin. The strong *GABA* signals of many amacrine cells virtually fill the inner plexiform layer and outline the characteristic bipolar cell terminals (Fig. 10F). GABA+ displaced amacrine cells that correspond to the glutamate-weak neurons of Figure 10C are indicated. The GABA signals in the amacrine cell layer are diverse and, as will be shown, constitute distinct superclasses of amacrine cells. The horizontal cell layer contains weakly γ+ somas that we presume are the homologues of H1 horizontal cells in other cyprinids. But at this scale and orientation, it is difficult to be certain of their identities. *Glutamine* signals are strongly differentiated (Fig. 10G): moderate in Müller cells, low in most bipolar cells, diverse in amacrine cells and uniformly high (≥10 mM) in ganglion cells. This is similar to the pattern observed in the goldfish (Marc *et al.*, 1995) and is roughly replicated in the mammalian retina (Kalloniatis *et al.*, 1996; Marc *et al.*, 1998b; Marc & Jones, 2002).

The *taurine* signals of the inner retina are almost identical to those of the goldfish, but seem better differentiated, perhaps because of the better lamination of the inner plexiform layer (Fig. 10H). Both horizontal cells and ganglion cells are virtually devoid of taurine, while bipolar cells and their terminals possess exceptionally high levels. Amacrine cells display intermediate values. As will be discussed later, this patterning demonstrates that there is no obvious association of taurine with any simple functional property such as glutamatergic transmission (both bipolar cells and ganglion cells are glutamatergic, but have opposite taurine signals) or mode of signal integration (both bipolar cells and horizontal cells are sustained, but have opposite taurine signals).

It is easy to forget that all of the foregoing data represent signals in the very same set of cells. The display of signal triplets more effectively captures these data correlations. A conventional γGE → *rgb* triplet reveals that γ+ and G+ amacrine cells appear to be completely distinct populations in the dark-adapted zebrafish retina (Fig. 11A), though this is not a quantitative demonstration of that likelihood. Similarly, the E+ bipolar and ganglion cells are set apart as dark blue. But this view is incomplete, as the molecular phenotypes displayed by bipolar and ganglion cells are quite distinct. The use of τQE → *rgb* mapping completely segregates the entire E+ cohort of the retina into cyan ganglion cells, pink-magenta bipolar cells, and red photoreceptors (Fig. 11B). Thus each distinct glutamatergic cell type in the zebrafish retina has its own molecular phenotype. The Müller cells display a yellowish-green signal that is characteristic of this glial cell type across all vertebrates (Kalloniatis *et al.*, 1994; Marc *et al.*, 1995; Kalloniatis *et al.*, 1996; Marc *et al.*, 1998b; Kalloniatis & Tomisich, 1999). The taurine signal is significantly weaker in zebrafish than other cyprinids and this will be discussed later. Müller cell placement is better visualized with *glutathione* signals in a γJE $\rightarrow$ *rgb* mapping (Fig. 11C), especially the enrichment of processes around the horizontal cells in the outer plexiform layer and at the external limiting membrane. Finally, the positioning of horizontal cells is best set off with a τ DE $\rightarrow$ *rgb* mapping (Fig. 11D) that shows a cyan band of cells between the $\tau+$ photoreceptor and bipolar cell layers, and even some fine hue differentiation among horizontal cell classes.

LAYER VIEWS OF THE INNER RETINA

Even large numbers of registered vertical sections are inadequate for better distinction of cell classes simply because of the great diversity of classes. Larger, contiguous samples become essential and it is then important to create databases from horizontal or shallow oblique serial sections. As described in the methods, each image must be acquired at an optimal resolution for light microscopy, *i.e.* <250 nm/pixel. Resolutions higher than that simply produce empty data bins by replicating blurred neighbors. Lower resolutions limit the precision of registration. Figure 12 displays a full mosaic of registered ADEGJQ$\tau\gamma$ images for serial oblique sections through the layers of the inner retina. These mosaics have three advantages: (1) they document the uniformity of signal quality over the array... quality unequaled by any other immunochemical method; (2) they provide sufficient obliquity to garner large samples of some patterned elements; (3) they retain precise laminar boundaries. Revisiting the basis signals allows clearer differentiation of cell classes associated with each signal and a better appreciation of their patterning. As before each signal is shown in its serial sequence in the data array (Fig. 12A alanine 01; 12B aspartate 02; 12C glutamate 03; 12D GABA 04; 12E glycine 05; 12F glutathione 06; 12G glutamine 07; 12H taurine 08). The cellular segregations of alanine signals can be better visualized in oblique sections, even though they still represent modest concentrations. The patterning suggests a preferential association with a diffuse set of amacrine cells, and a characteristic banding pattern in the inner plexiform layer resembles a hybrid of glycine and GABA banding. The aspartate signals are complex, with patches of high signals in the horizontal, bipolar and amacrine cell layers, multiple laminae in the inner plexiform layer and distinct labeling of ON-center bipolar cell terminals. The glutamate distribution is likewise complex but more interpretable, based on previous data from goldfish: the band in the outer plexiform layer is mostly composed of bipolar cell dendrites, with an interposed wide band of horizontal cells, followed by the strong glutamate signals of the bipolar cell layer proper. The amacrine cell layer displays characteristic mixed, weak glutamate signals that we shall explore in more detail shortly. Showing more clearly than in vertical sections, the highest glycine signals are localized to a poorly laminated set of amacrine cells distributed throughout the amacrine cell layer, slightly biased to the distal half. A band of strong glycine signals, unique to the zebrafish among cyprinids, is positioned at roughly level 50 of the inner plexiform layer.

Selected *rgb* triplets preferentially reveal subsets of the entire retinal cell cohort, even at low magnification. The subsets are overlapping, but this display is not merely an academic exercise as the correct triplets can form a screen for specific cell populations, especially in disease states (Marc *et al.*, 1998a). The γGE triplet (Fig. 13A) is a standard that captures all neurons in the retina and categorizes them as in a single image. A sprinkling of cyan cells in the inner nuclear layer represents the sparse glycinergic interplexiform cell population. A decrease in GABA signal occurs at about level 50 of the inner plexiform layer, where a class of G+ amacrine cell has a dense arborization, but the overall GABA signal is otherwise uniform throughout, as in the goldfish (Marc & Liu, 2000). The τQE triplet (Fig. 13B) is diagnostic of Müller cells due to their extremely high glutamine contents. However, the taurine content of Müller cells in the zebrafish is roughly half that found in the goldfish and other vertebrates, giving the overall image a green cast. The bipolar cell layer is distinctively magenta, as in all vertebrates, while the photoreceptor layer is red-shifted, indicating the much higher taurine content in photoreceptors. Though difficult to see at low magnification, the γJA triplet (Fig. 13C) isolates the extensive Müller cell branching in the outer plexiform layer better than any other and also indicates the partial segregation of alanine-rich cells from the $\gamma+$ amacrine cell cohort in the inner nuclear layer. The array of black dots in the middle of the image, deep in the inner plexiform layer, are the terminal arrays of Bb ON-center bipolar cells. Conversely, a $\gamma\tau$D mapping (Fig. 13D) exposes the ganglion cells as black dots deep in the inner plexiform layer, whereas most bipolar cell terminals become a cyan hue. Mixed blue and red signals in the outer plexiform layer denote multiple horizontal cell types. One of the richest triplets for displaying subsets of cell types is the γDE mapping (Fig. 13E), with distinct cyan rod horizontal cells in the horizontal cell layer, pale blue aspartate-rich ON-center bipolar cells, mixed GABA amacrine cells cohorts of red, orange and yellow. This metabolic diversity serves as a marker for cell classes, even though we have no biochemical model to explain it. The complex banding of the inner plexiform layer is particularly rich, and a discerning trichromat can likely pick out at least eight layers. We will return to this point later, as it is a key issue in the characterization of the zebrafish retina. Finally, the AGJ mapping (Fig. 13F) is subtle, but shows that most of the high alanine contents of cells in the zebrafish retina (and in mammals, as well) are associated with G+ amacrine

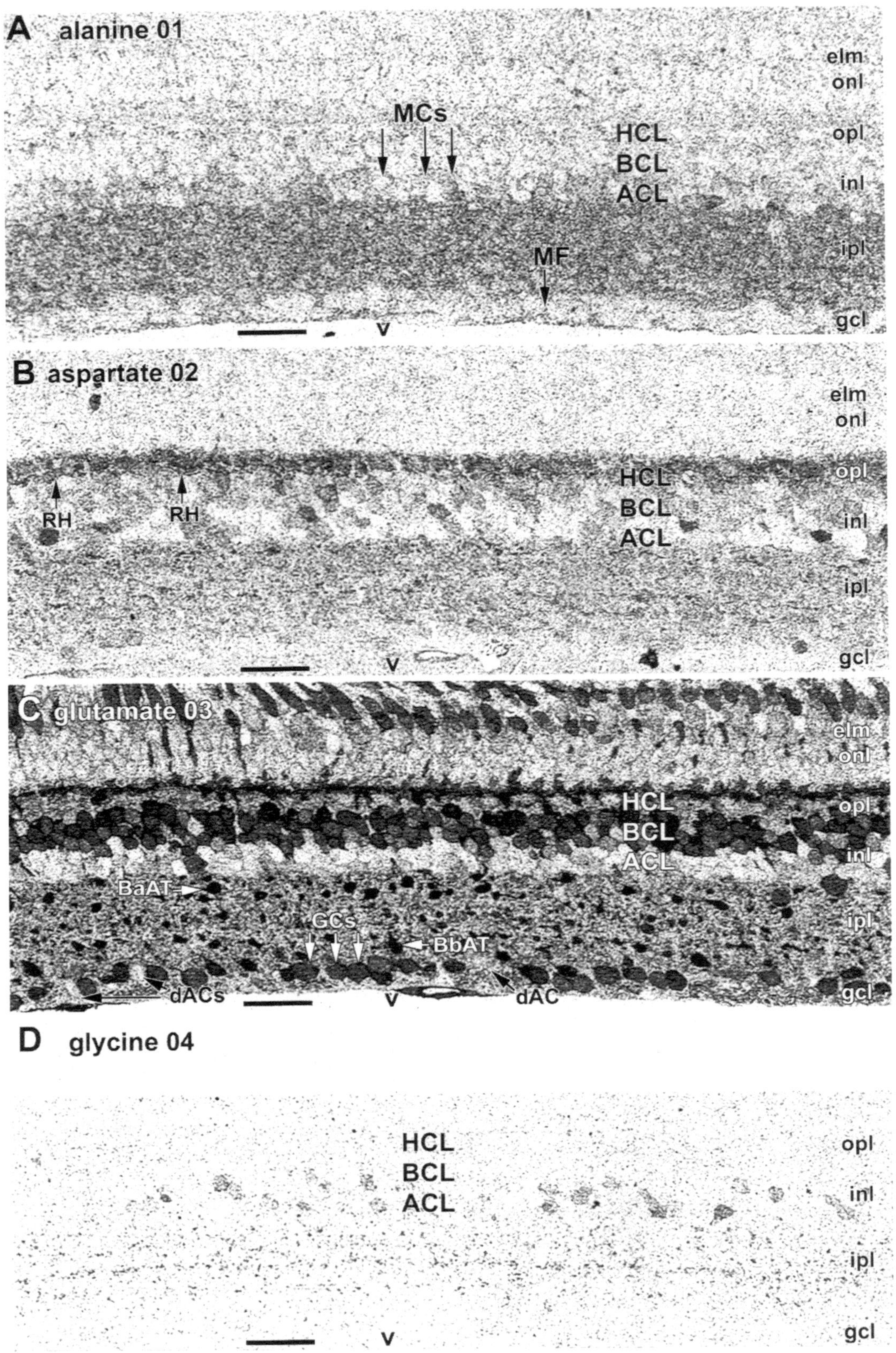
A alanine 01
elm
onl
MCs
HCL
opl
BCL
ACL
inl
ipl
MF
gcl
v
B aspartate 02
elm
onl
opl
HCL
RH
RH
BCL
inl
ACL
ipl
gcl
v
C glutamate 03
elm
onl
HCL
opl
BCL
inl
ACL
BaAT
ipl
GCs
BbAT
dACs
v
dAC
gcl
D glycine 04
HCL
opl
BCL
inl
ACL
ipl
gcl
v

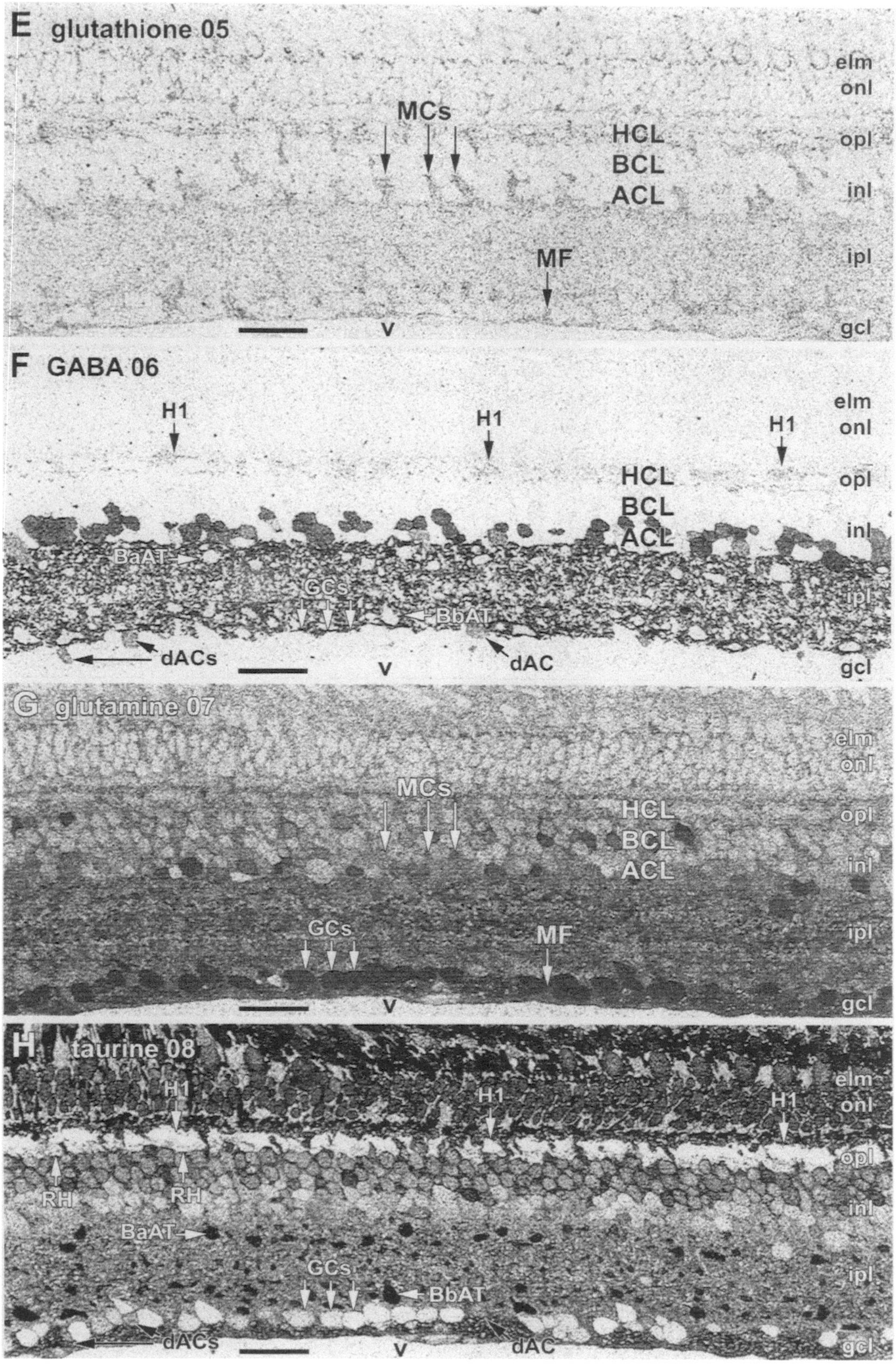

Fig. 10. Single channel, registered ADEGJQτγ signals in the dark-adapted inner retina. A, alanine 01. B, aspartate 02. C, glutamate 03. D, glycine 04. E, glutathione 05. F, GABA 06. G, glutamine 07. H, taurine 08. Each channel highlights a different set of cellular compartments. Section thickness: 250 nm. Scale 20 μm. Abbreviations: inl, inner nuclear layer; MC, Müller cells; RH, rod horizontal cells; dAC, displaced amacrine cells; other abbreviations as in Figures 3–5.

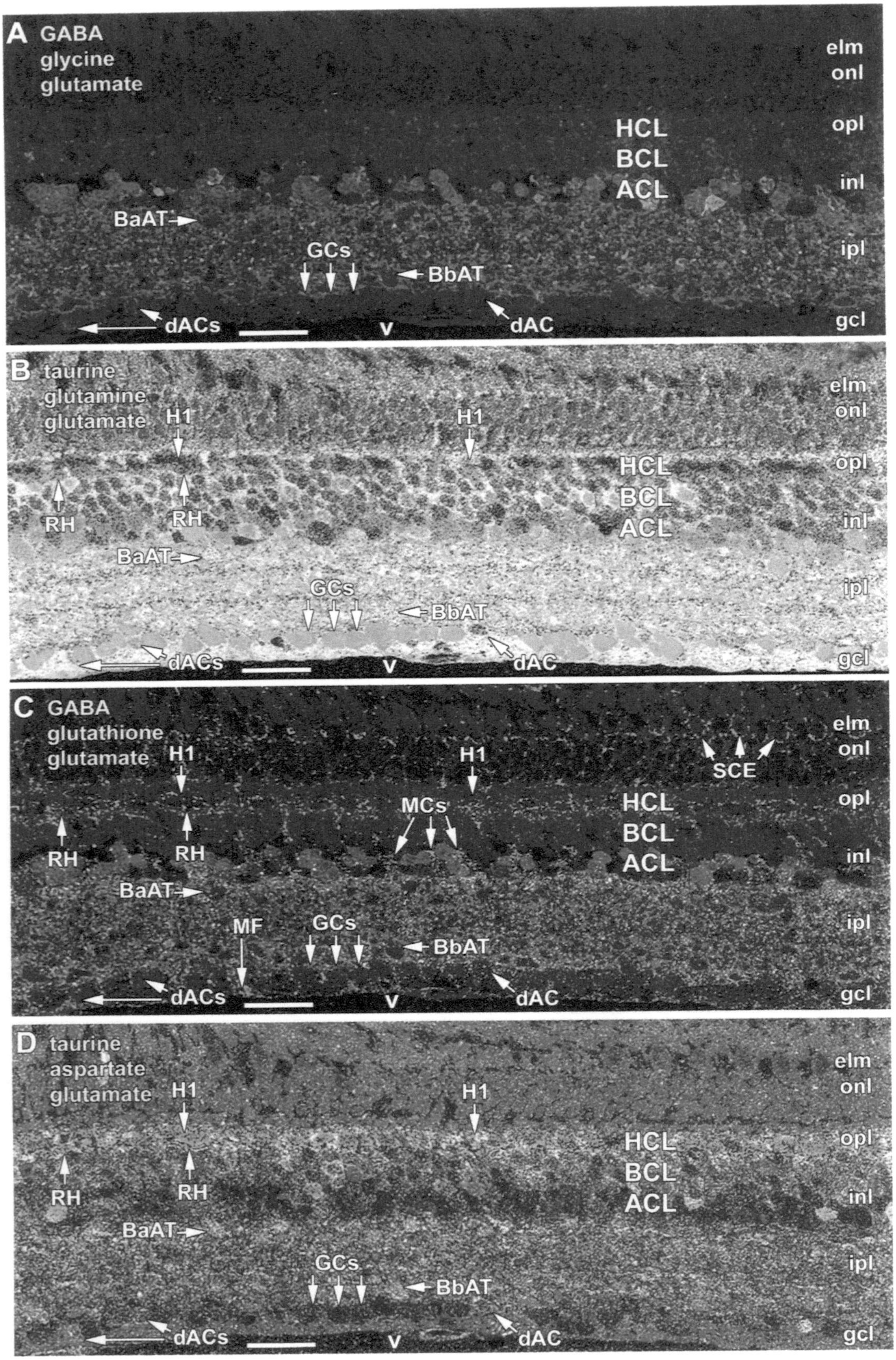

Fig. 11. Triplet mapped basis set signals in the dark-adapted inner retina. A, γGE $\rightarrow$ *rgb*. B, τQE $\rightarrow$ *rgb*. C, γJE $\rightarrow$ *rgb*. D, τDE $\rightarrow$ *rgb*. Each triplet delineates a subset of elements in the retina, but no one triplet uniquely captures all elements.

cells, including the characteristic band at level 50 of the inner plexiform layer.

CLASSIFICATION OF THE INNER NUCLEAR LAYER

The comprehensive dataset, with its rich differentiations of cell cohorts, allows the luxuries of a large sample and high resolution. One can literally zoom into these datasets and browse cell populations at optimum resolution (Fig. 14A–H). This oblique sampling permits high cell numbers and preservation of laminar boundaries. Even casual observation allows numerous distinctions of possible cell classes and metabolic associations. For example, there is a strong albeit incomplete correlation between alanine (Fig. 14A) signals and glycine (Fig. 14E) signals; this is the first metabolic clue regarding the possible mechanism of glycine synthesis in glycinergic amacrine cells (see Discussion). Aspartate (Fig. 14B) signals definitely reveal metabolic subsets of horizontal, bipolar and amacrine cell populations. As in every retina, the glutamate (Fig. 14C) contents of bipolar cells are variable and we might well ask whether this is simply metabolic instability or specific metabolic profiles expressed by bipolar cell classes. The fact that some cones change their glutamate profiles in adaptation suggests that many cells could alter profiles for many physiological reasons, suggesting we should proceed with care; that we perhaps ought not equate metabolic and natural classes without other data. Classification can help this analysis, as we will see later. The diversity in GABA (Fig. 14D) signal strength is similar: some cells have high signals, others low, and it is obvious that some somas in the bipolar cell layer have elevated GABA signals. The origin and significance of these variations, in the weak signals especially, need to be sorted out, but this is not possible by surveying monochrome images alone.

An important feature emerges in these images: the layer of fine, γ+ processes immediately proximal to the H1 horizontal cell somas. These are the axon terminal expansions of the cone horizontal cells. We will later discuss this point in detail for those unfamiliar with this teleostean neuronal oddity, but for now simply note that these processes are much smaller ($\approx 1\ \mu$m in width, unknown length) than those in the closely related carp and goldfish (5–10 μm in width, up to 400 μm long). In general, only Müller cells show high levels of glutathione (Fig. 14F) signals, so glutathione is a good Müller cell marker in the zebrafish. Glutamine (Fig. 14G) levels help segregate Müller cells to some degree, but oblique sections reveal significant variation in glutamine signals across both amacrine and bipolar cells. Finally, the negative taurine (Fig. 14H) signals of horizontal cells, however inexplicable, are a stroke of good fortune, as they cleanly separate them from all other cell classes in a binary fashion. The rich mesh of horizontal axon terminal expansions is particularly evident, suggesting that there are many more kinds of cone horizontal cells than γ+ H1 horizontal cells.

Subsets of cell types become even more evident in triplet mappings. The standard γGE mapping (Fig. 15A) displays the complete quantitative segregation of γ+ and G+ amacrine cells, including glycinergic interplexiform cells. And amidst the deep blue bipolar cell layer array, one can detect the faint red γ+ bipolar cells. In the horizontal cell layer proper, both H1 horizontal cell somas and their axon terminals evince a weak γ+ signal. In Figure 15B, we have used a mixed τQE → **c***gb* mapping that strongly segregates the Müller cells, horizontal cells and their axon terminals. Most of the Müller cell somas are positioned at the interface between the amacrine and bipolar cell layers. Figure 15C is also a mixed γDE → **c***g***y** mapping that simultaneously discriminates rod horizontal cells, two subsets of the bipolar cell cohort, and several subsets of the γ+ amacrine cell population. Finally, AGJ → *rgb* mapping (Fig. 15D) clearly links the highest alanine contents with the highest glycine contents in the amacrine cell layer and the glycinergic interplexiform cells (Marc & Lam, 1981; Marc & Liu, 1984).

Given this complex array of signals, one might reasonably despair of further resolving all the cell types. This is where molecular phenotyping, clustering and deconvolution analyses come into play. Molecular phenotyping effectively covers all cellular space: no cells are left out of the analysis. Figure 16 is the final map produced by assigning the proper molecular phenotype class code to every cell in the image. The γ+ amacrine cell type was readily divisible into seven molecular phenotype classes. This does not mean that there are only seven natural classes of γ+ amacrine cells in the zebrafish: only that there are seven stable molecular phenotype groupings, each likely to represent a superclass of many natural classes sharing the same signature. The G+ amacrine type was coded as a single superclass for simplicity, but deconvolution analysis suggests that it is composed of no less than three groups. Unlike other cyprinids analyzed so far, the zebrafish has remarkably strong metabolic differentiation across bipolar cells, and six classes were resolved. Some of these are certainly natural classes, as will be discussed, while most are superclasses of many natural classes. Finally, the horizontal cell layer is composed of islands of proximal cells that we presume to be rod horizontal cells, intermediate islands likely to be Hx cone horizontal cells and distal γ+ H1 horizontal cells. We will consider the exact signatures underlying each of these cell classes below.

HORIZONTAL CELLS

Knowing that the inner nuclear layer can be formally classified by molecular phenotyping allows a more

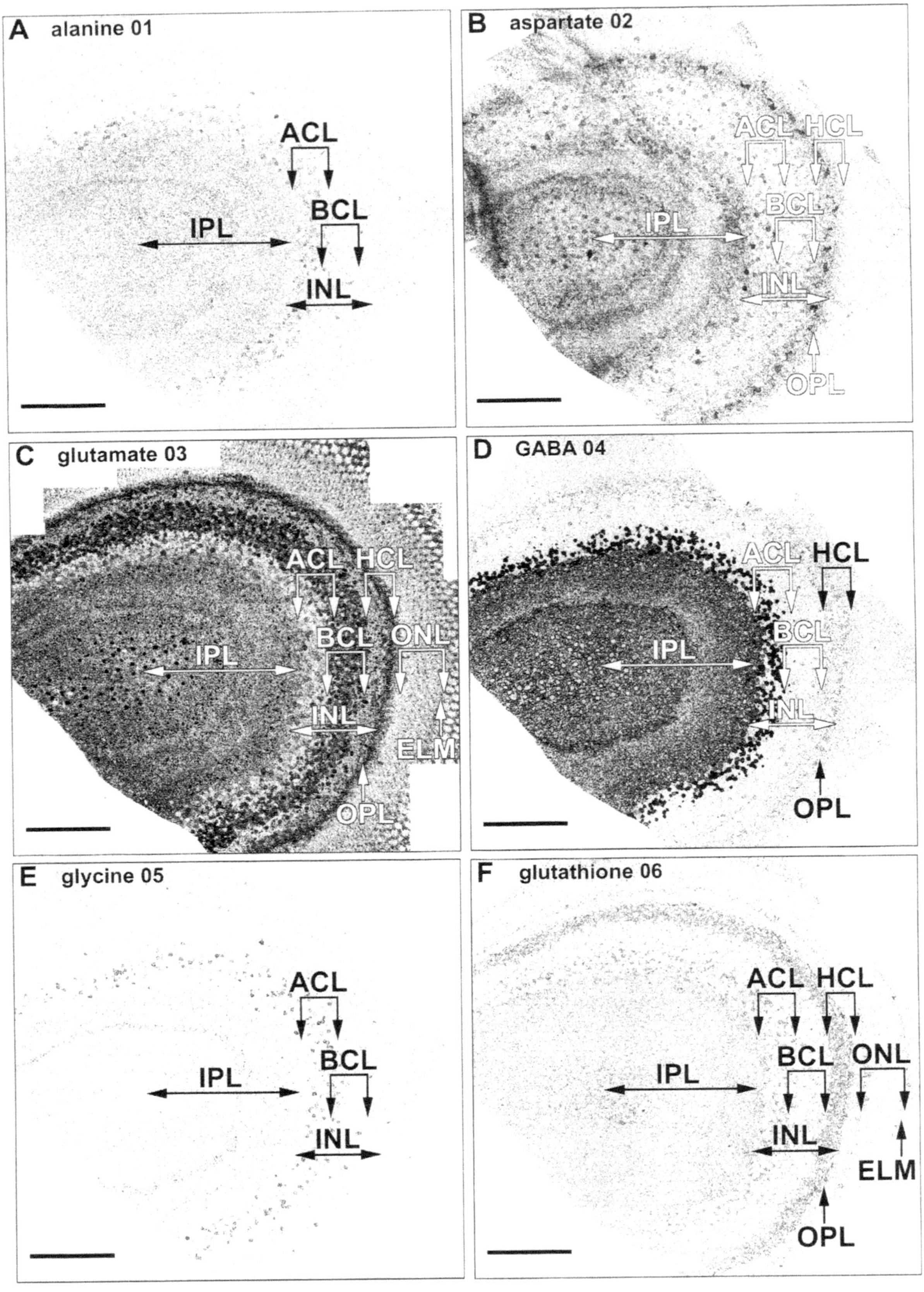
A alanine 01
ACL
BCL
IPL
INL
B aspartate 02
ACL HCL
BCL
IPL
INL
OPL
C glutamate 03
ACL HCL
BCL ONL
IPL
INL
ELM
OPL
D GABA 04
ACL HCL
BCL
IPL
INL
OPL
E glycine 05
ACL
BCL
IPL
INL
F glutathione 06
ACL HCL
BCL ONL
IPL
INL
ELM
OPL

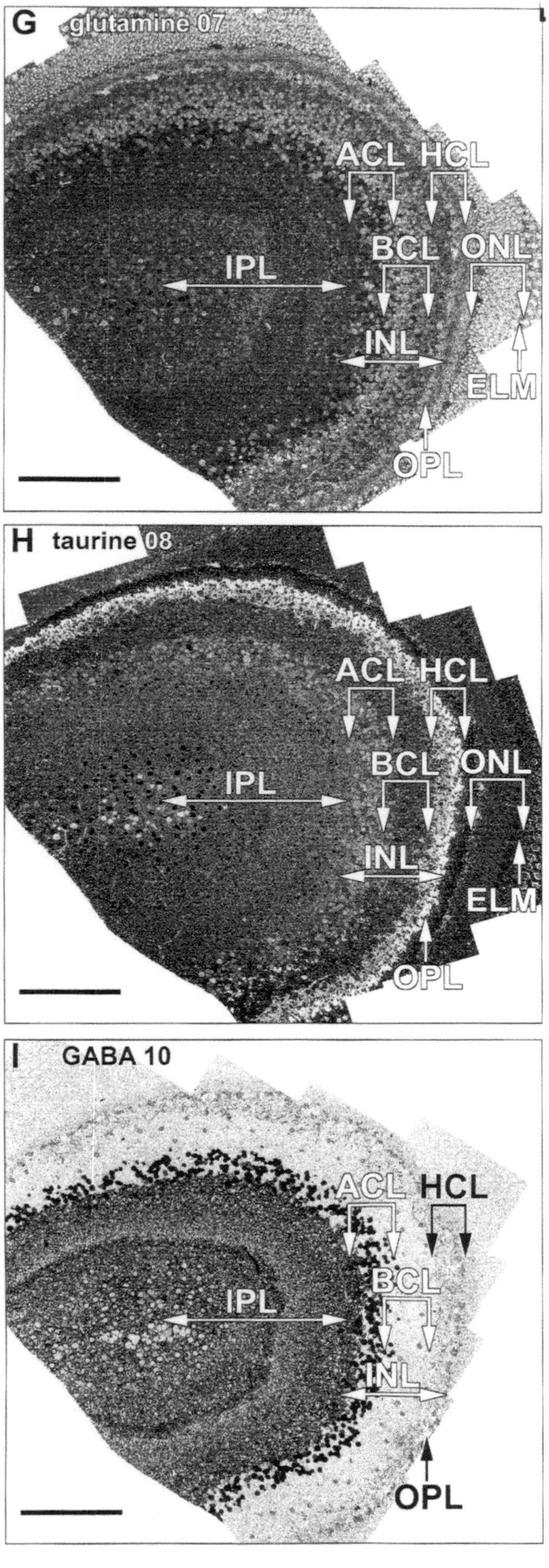

aggressive decomposition of cell classes in each major cell type. How many horizontal cell classes are there? This is not a trivial question since we know that all other cyprinids studied structurally and/or physiologically display four types: spectrally monophasic, $\gamma+$ H1 cells, biphasic $\gamma-$ H2 cells, triphasic $\gamma-$ H3 cells, and $\gamma-$ rod horizontal cells. The photoreceptor types of the zebrafish retina are chromatically sufficient to drive four such horizontal cell types. By combining serial trangential, registered GABA (Fig. 17A), aspartate (Fig. 17B) and taurine (Fig. 17C) signals, we can get a sense of the relative population sizes of the three horizontal cells groups in the zebrafish. A doublet mapping of $\gamma \rightarrow$ red, $\tau \rightarrow$ blue, combined with an aspartate channel displayed as an edge filtered yellow overlay (Fig. 18A) shows that the distal layer of H1 horizontal cells intermeshes tightly with numerous $\gamma-/\tau-$ Hx horizontal cells, but is well separated from the D+ presumed rod horizontal cells that form the proximal margin of the horizontal cell layer. Recoded into a theme map (Fig. 18B), these data allow estimates of cell densities. The rod horizontal cells tile the retina at $\approx 2200/\text{mm}^2$, while the H1 horizontal cells are much denser at $\approx 5600/\text{mm}^2$. This is much higher than the density of H1 horizontal cells in the goldfish, but it is consistent with the fact that the soma sizes of zebrafish horizontal cells are smaller than those of other cyprinids by a factor of 2. The density of Hx cells is more difficult to estimate as individual somas do not cleanly emerge from the dataset. Even so, they take up twice the area of H1 horizontal cells. If the classes of horizontal cells scale with size and cone density as in other cyprinids, this would argue that the Hx group is a composite of natural classes and that the zebrafish retina likely contains all 4 known classes of cyprinid horizontal cells.

BIPOLAR CELLS

The metabolic profiles of bipolar cells are more complex than in any other cyprinid and, while we have no explanation for this, it does permit better discrimination of bipolar cell superclasses. Subsets of mildly D+ cells (Fig. 19A) reside within the mix of cells with varying glutamate signals (Fig. 19B). Unique to the zebrafish, among cyprinids at least, is a distinct population of presumed bipolar cells with moderate GABA signals (Fig. 19C). Another subset with very weak but distinct GABA signals (≈ 0.2 mM) is interspersed, but these are also evident in the goldfish retina (Marc, unpublished

Fig. 12. A high-resolution library of registered oblique mosaic channels: ADEGJQ$\tau\gamma$ signals in the dark-adapted retina. A, alanine 01. B, aspartate 02. C, glutamate 03. D, GABA 04. E, glycine 05. F, glutathione 06. G, glutamine 07. H, taurine 08. I, GABA 10. Section thickness: 250 nm. Scale 50 μm.

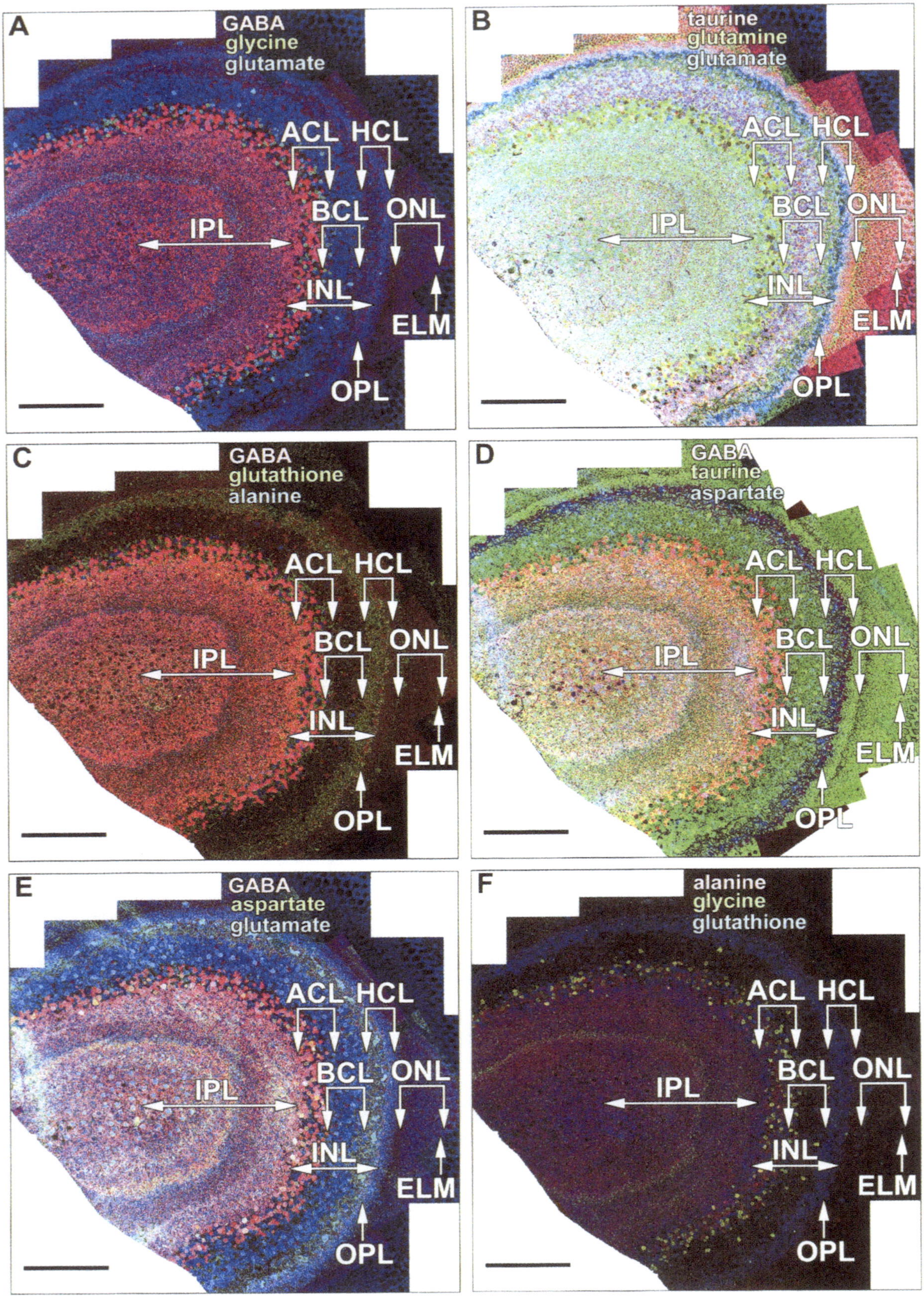

Fig. 13. Triplet mappings of the high-resolution library: ADEGJQ$\tau\gamma$ signals in the dark-adapted retina. A, γGE $\rightarrow$ *rgb*. B, τQE $\rightarrow$ *rgb*. C, γJA $\rightarrow$ *rgb*. D, $\gamma\tau$D $\rightarrow$ *rgb*. E, γDE $\rightarrow$ *rgb*. F, AGJ $\rightarrow$ *rgb*. Scale 50 μm.

data) and we will consider their identities when the terminal arrays of ON-center bipolar cells are described. Some of the cellular space at the proximal margin of the bipolar cell layer is comprised of Müller cell cytoplasm (Fig. 19D). Both glutamine (Fig. 19E) and taurine (Fig. 19F) are extremely varied across bipolar cells, the glutamine signals being weaker and the taurine signals stronger than those of amacrine, ganglion and Müller cells. As revealed by classification and triplet mapping, the γ+ and D+ sets are completely distinct populations in γDE → *rgb* (Fig. 20A), τQJ → *rgb* (Fig. 20B), and γDJ → *rgb* space (Fig. 20C). The τQJ mapping in particular is important because it captures all cells in the array, and though τQJ signals are not the primary elements of the differentiation, the γ+ bipolar cells are sufficiently Q− to emerge from the complex display. The γ+ cells are also very low in glutamate content and, on balance, have molecular phenotypes more reminiscent of H1 horizontal cells (but for the elevated taurine signal) than a classical glutamatergic bipolar cell.

THE AMACRINE CELL LAYER

The real challenge in classification is the amacrine cell layer: the most diverse groups of cells in any retina. This is particularly critical in the cyprinid retina. The roach for example likely has >70 amacrine cell classes (Wagner & Wagner, 1988). Oblique sections through the amacrine cell layer give us a detailed view of the likely molecular phenotypes involved. Figure 21 shows the registered basis set of amacrine cell signals in the dark-adapted zebrafish retina. This provides the most revealing of all views of the distribution of alanine signals (Fig. 21A) showing that it is extremely heterogeneous: many cells maintain no detectable alanine levels (<50 μM), while various cells approach 1 mM. The distribution of aspartate signals (Fig. 21B) shows a greater range, but a more restricted distribution, whereas glutamate signals (Fig. 21C) are extremely complex. Most cells contain some glutamate, but rarely at the levels expressed by bipolar cells. The GABA signals (Fig. 21D) also reveal substantial variation in GABA content across cells. Is this simply metabolic instability within a class, or a metabolic signature across different classes? As with the bipolar cell cohort, classification will provide some evidence that these are stable classes. The G+ amacrine cells (Fig. 21E) are clearly sparser and smaller than γ+ amacrine cells and partly correlate with the alanine distribution. As in vertical sections, glutathione (Fig. 21F) is consistently and preferentially associated with the Müller cell population in the dark-adapted retina. Both glutamine (Fig. 21G) and taurine (Fig. 21H) display extremely complex distributions within the amacrine cell cohort. Alone, they are too difficult to sort, but triplet mapping and classification shows distinct patterns underlying this complexity. In a triplet mapping of γGE → *rgb*, 215 γ+ amacrine cells are completely segregated from 126 G+ amacrine cells (Fig. 22A). This determination is possible only through quantitative image registration. Aspartate enrichment is exclusive to a subset of the γ+ amacrine cell cohort (Fig. 22B, γDE → *rgb*). Most amacrine cells contain a significant amount of taurine, although less than most bipolar cells and more than all ganglion or horizontal cells, leading to a complex display of τJE mixtures (Fig. 22C). Roughly half of all amacrine cells have elevated alanine signals, and virtually all glycine cells map onto that cohort (Fig. 22D.)

This complexity is partly resolved by classification of molecular phenotypes into a theme map (Fig. 23). Again, seven classes of γ+ amacrine cells emerge. While it is beyond the scope of this manuscript to also perform a complete morphometric analysis of all these classes, we can assert that the different classes also have distinctive sizes: *e.g.* classes 1 and 2 have smaller mean diameters than classes 3 and 4. Our molecular classification is not informed by structure, and the fact that it discovers cell size groups is an independent proof that the metabolic profiles have a stable biologic origin, and are not just random associations (see Marc & Jones (2002)). This sample of the amacrine cell layer also displays enough glycinergic cells to capture a few cells with exceptionally glycine-weak signals, labeled here as class 2. The two unknown γ− G− cells are likely dopaminergic interplexiform cells, as in goldfish (Jones *et al.*, 1999).

THE INNER PLEXIFORM LAYER

The single most dramatic feature of retinal organization is perhaps the functional stratification or lamination of the inner plexiform layer, which is particularly complex in cyprinids (Marc, 1986; Wagner & Wagner, 1988). Figure 24 summarizes the major layers of the zebrafish inner plexiform layer as reflected in the distribution of bipolar cell axon terminals and the positioning of complex signal borders (*e.g.* Fig. 13). As in all cyprinids, the distribution of GABA signals is nearly uniform but for a small mid-retinal dip (Fig. 13A; see also Marc & Liu (2000)). Previous publications suggesting multibanded GABA signals in the inner plexiform layer, with interspersed GABA-free zones, are corrupted by postmortem ischemic release, which is particularly rapid and extreme in the zebrafish. There are no fewer than 12 distinct laminae in the zebrafish inner plexiform layer and, by superposition of additional signals, many more are likely to emerge. The Bb layer between layers 90–100 contains terminals that belong to the homologues of cyprinid mixed rod-cone ON-center bipolar cells, and likely all three are present (Bb1, Bb2, Bb3). The distal border of the inner plexiform layer is a bipolar cell-free gap followed by a broad zone divisible into areas where homologues of two types of cyprinid mixed rod-cone OFF-center bipolar cells terminate (see Marc *et al.*

(1988)). Mixed within and between these specialized zones are the characteristic small terminals of pure cone bipolar cells (Sherry & Yazulla, 1992). Specific size groupings (*e.g.*, C *fine*, C *medium*) are segregated from other bipolar cell terminals bands by three gaps of relative bipolar cell process depletion. These internal gaps are not bipolar cell process-free, but reduced in terminal density. The lower half of layer Ba2 and gap3 are the zones occupied by dendrites of cholinergic amacrine cells in goldfish retina (Tumosa *et al.*, 1984; Marc, 1986; Tumosa & Stell, 1986; Marc & Liu, 2000) and the lower half of layer C *fine* is where the distinctive level 50 G+ amacrine cells terminate. Part of our objective is to provide as much discrimination of bipolar cell types as possible with the basis set of signals. As noted above, a subset of bipolar cells maintains an elevated level of aspartate and this is also expressed in their axon terminals (Fig. 25A). These same bipolar cells are also revealed

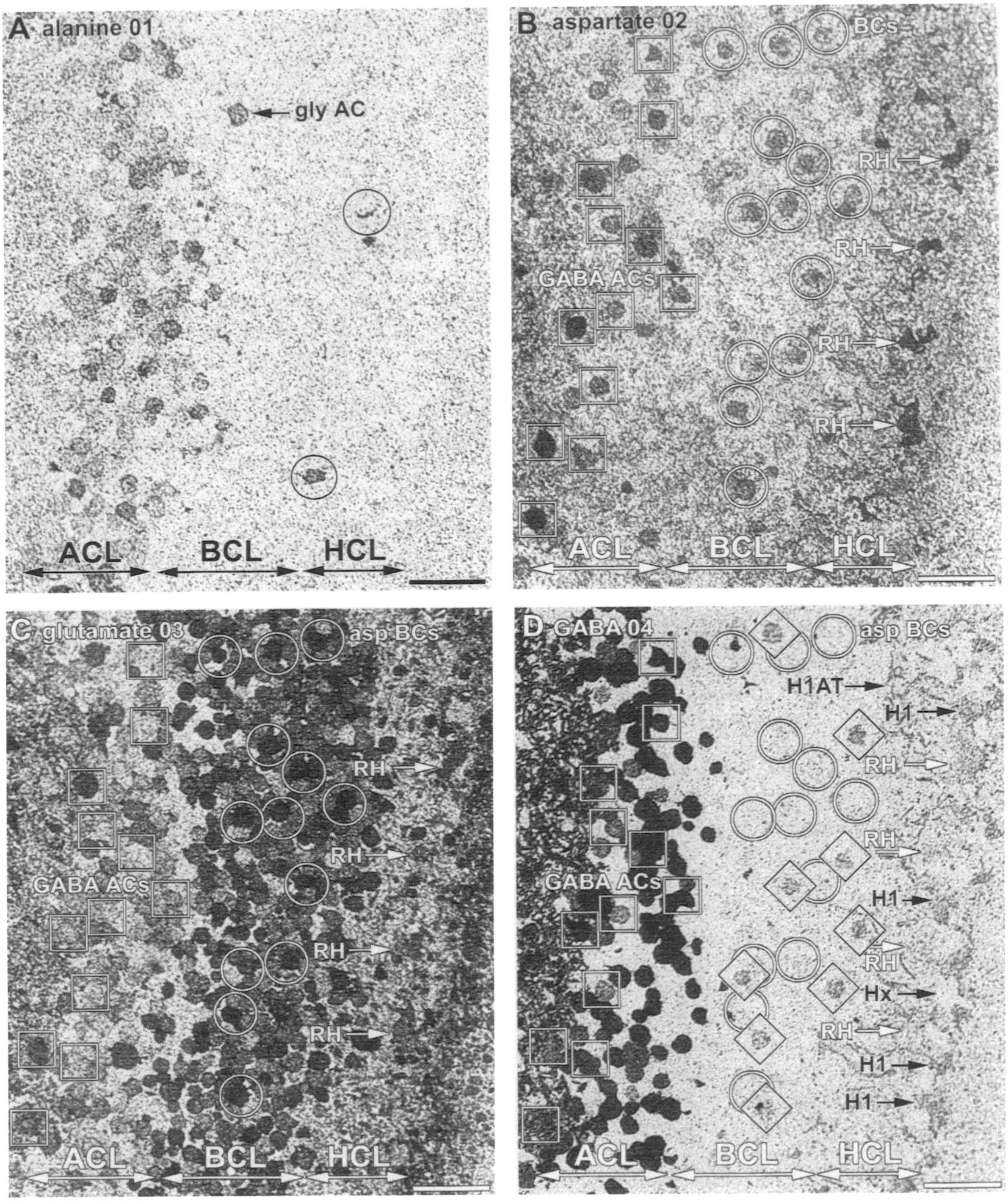

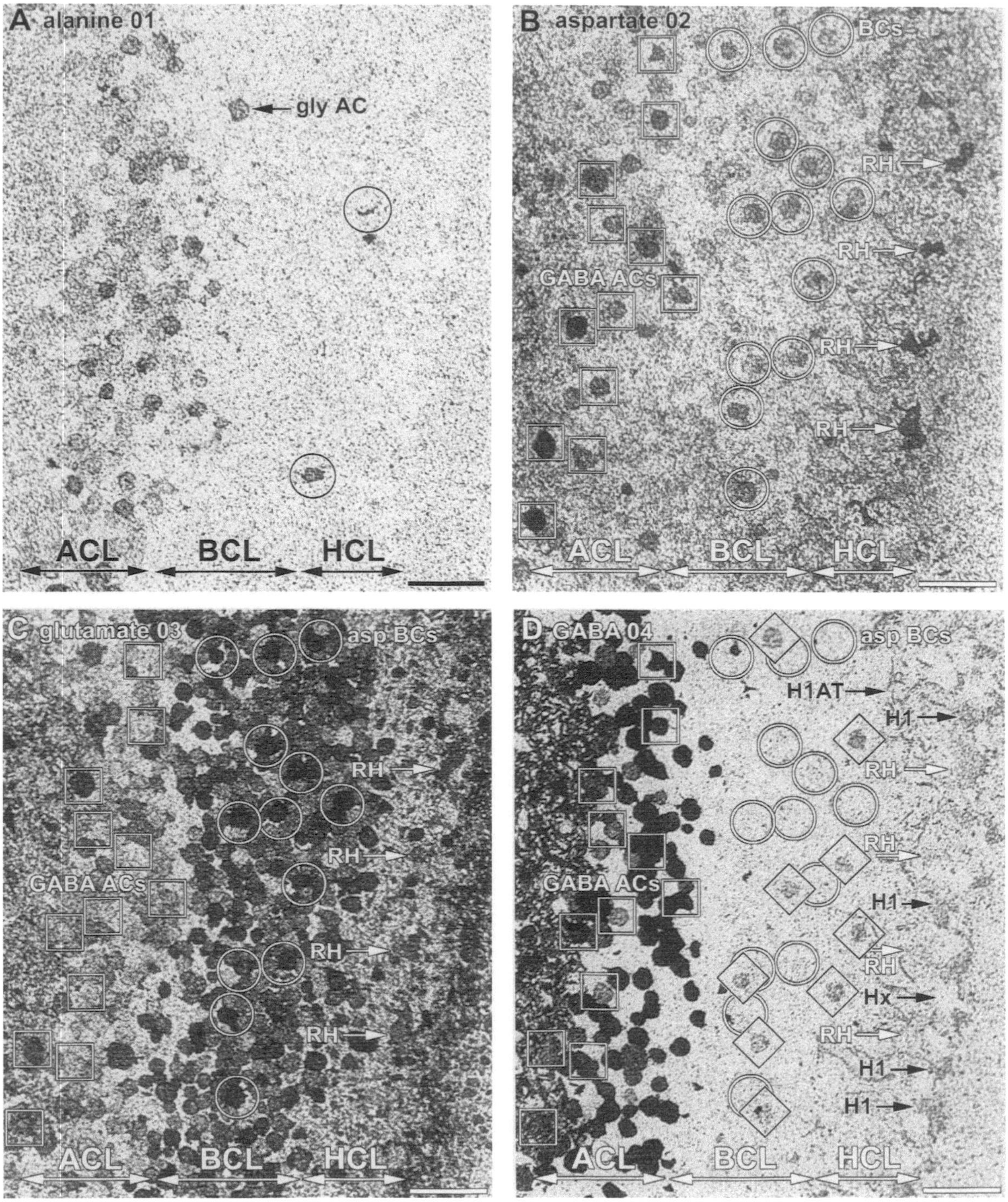

Fig. 14. A high-resolution library of the inner nuclear layer: ADEGJQ$\tau\gamma$ signals in the dark-adapted retina. A, alanine 01. Alanine-rich glycinergic interplexiform cells are circled. B, aspartate 02. Aspartate-rich bipolar cells are circled and a mélange of aspartate+ amacrine cells are boxed. C, glutamate 03. Bipolar and amacrine cells are marked as in panel B. D, GABA 04. Bipolar and amacrine cells are marked as in panel B. GABA+ presumed bipolar cells are marked by diamonds. Both H1 and Hx horizontal cells are indicated, as well a fine, strand-like horizontal cell axon terminals (H1AT). E, glycine 05. Glycinergic interplexiform cells are circled. F, glutathione 06. Large, glutathione-deficient bipolar cell somas that protrude into the horizontal cell layer are circled. G, glutamine 07. H, taurine 08. Section thickness: 250 nm. Scale 20 μm.

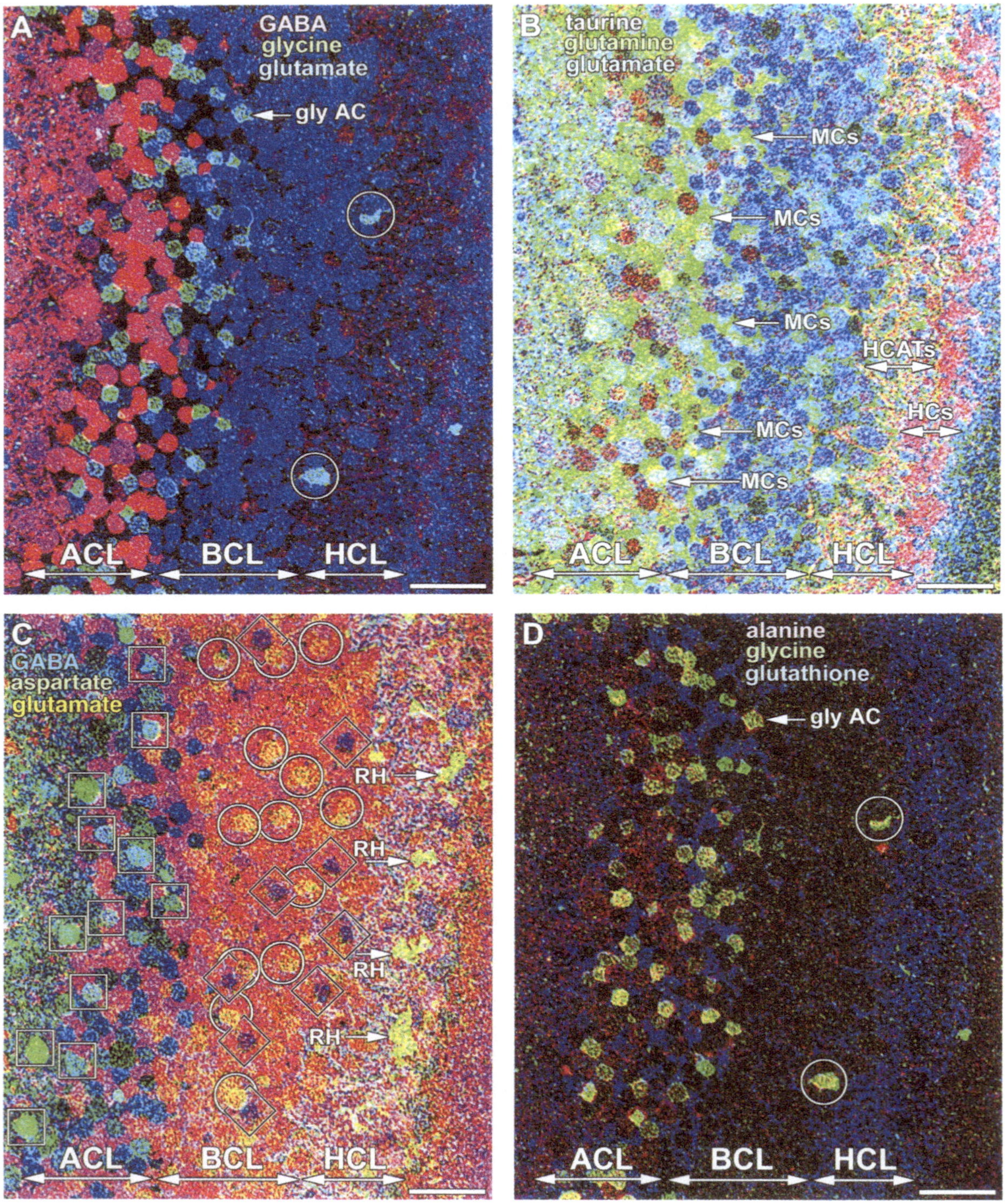

Fig. 15. Triplet mappings of a high-resolution library of the inner nuclear layer. A, γGE $\rightarrow$ *rgb*. B, τQE $\rightarrow$ *rgb*. C, γJA $\rightarrow$ **c***g***y** complex mapping. D, AGJ $\rightarrow$ *rgb*. Abbreviations and symbols as in Figure 14. Scale 20 μm.

by glutamate, GABA and taurine signals (Fig. 25B–D) but with subtle differences. The array can be visualized as a conventional γDτ $\rightarrow$ *rgb* mapping and enhanced in a γDτ $\rightarrow$ **cmy** mapping (Fig. 26A and B). The key result is that a subset of terminals shows a slightly elevated GABA content, as do some cell bodies in the outer nuclear layer (*e.g.* Fig. 19C), and they comprise a well-patterned array of irregular, rather than circular terminal profiles. This subset is likely to be class Bb3 (Ishida *et al.*, 1980b) and has a density of 1300/mm^2:

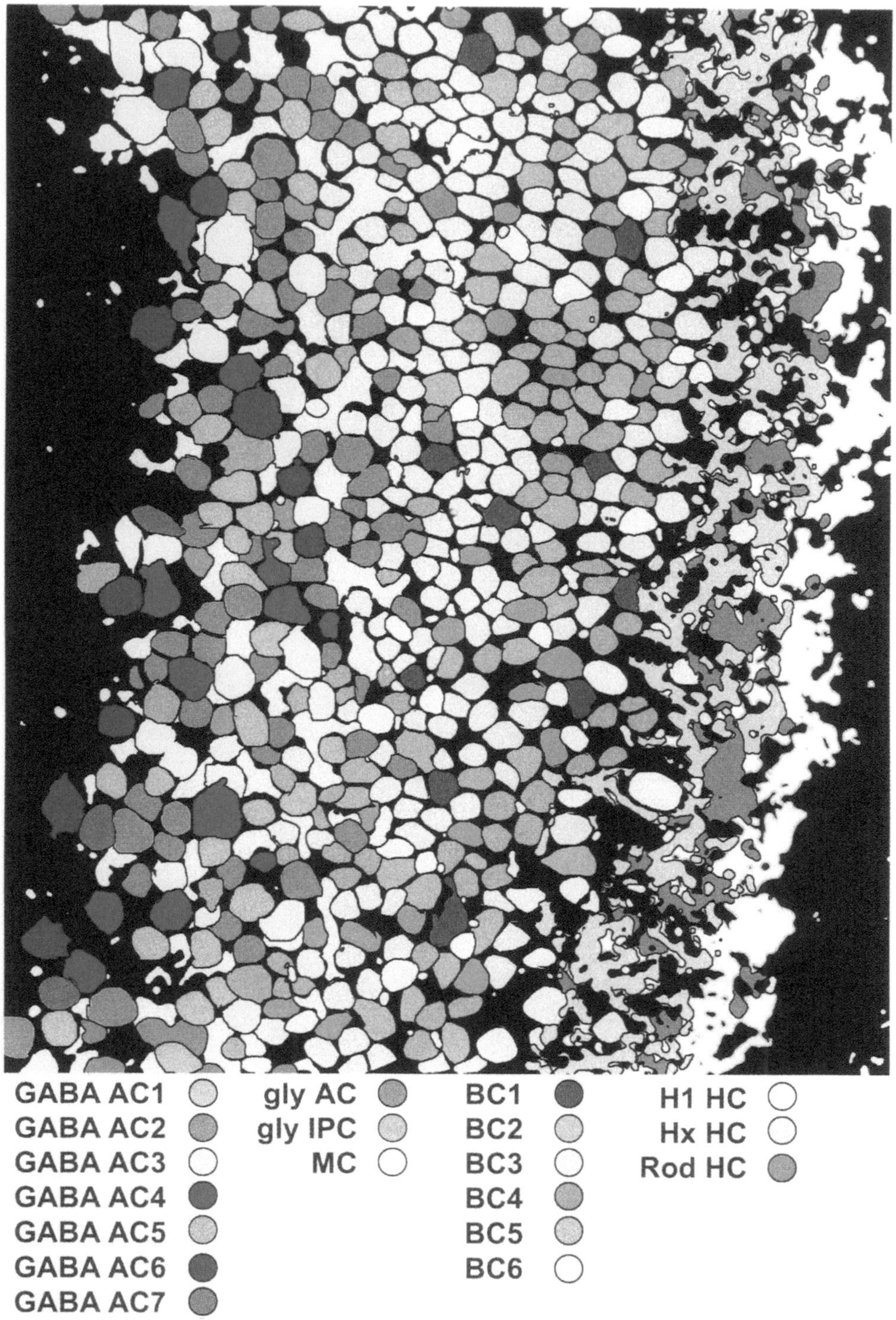

Fig. 16. A formal theme map of the dark-adapted inner nuclear layer. Each cell type is color coded by class membership based on isodata clustering and histogram deconvolution. The precise stratification of the inner nuclear layer into horizontal, bipolar, Müller and amacrine cell layers in the theme map is derived purely from the basis set (chemical space) and not morphological indices.

it corresponds to class BC6 in the theme map of Figure 16. The remaining arrays are likely a mixture of structural classes Bb1 and Bb2 ($3700/mm^2$) submerged in the molecular superclasses BC4 or BC5.

THE GANGLION CELL LAYER

The ganglion cell layer is a monolayer just beneath the inner plexiform layer and harbors a mixture of ganglion cell classes and displaced amacrine cells. All

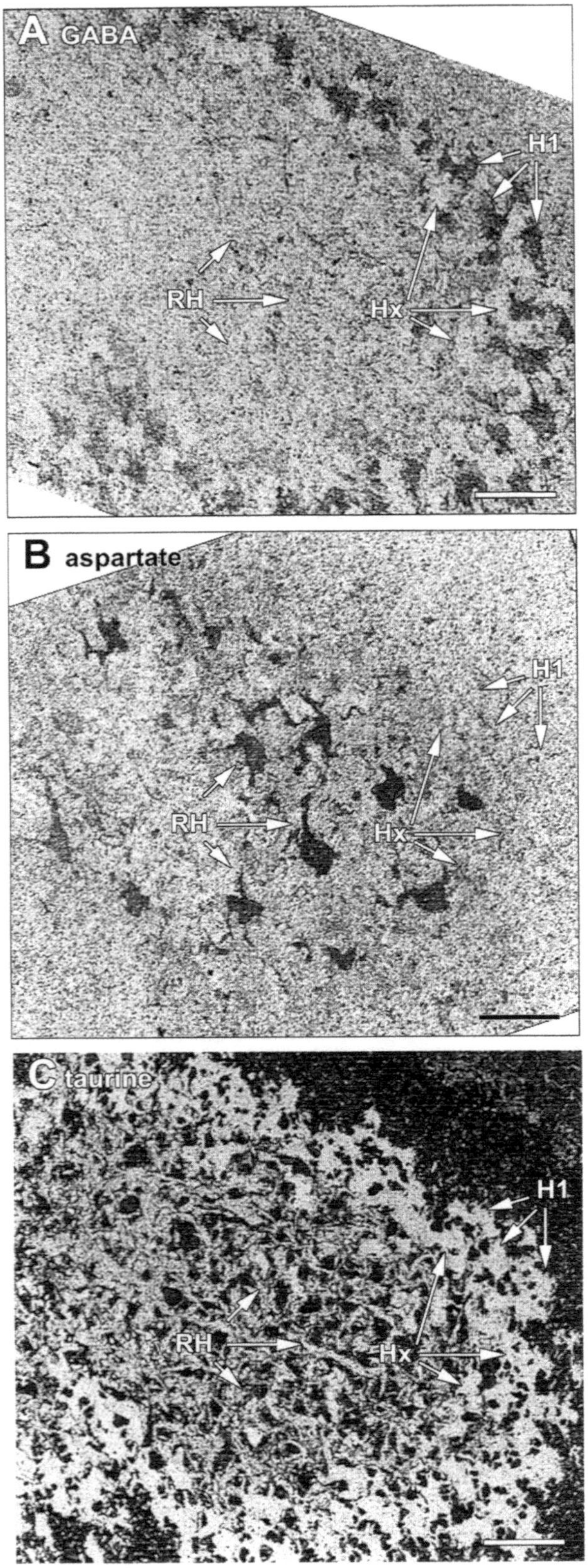

Fig. 17. A high-resolution library of the horizontal cell layer: $\gamma D\tau$ signals in the dark-adapted retina. A, GABA. B, aspartate. C, taurine, 03. Section thickness: 250 nm. Scale 20 μm.

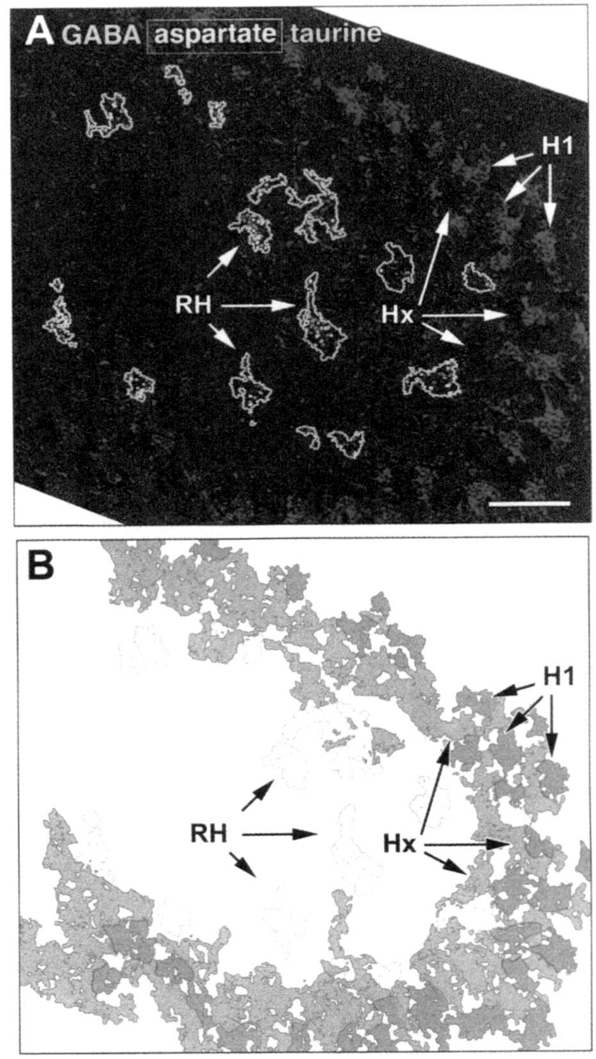

Fig. 18. A composite mapping of the horizontal cell layer. A. $\gamma\tau \rightarrow rb$ signals in the dark-adapted retina with edge-filtered D+ rod horizontal cells superimposed in yellow. B, theme map of horizontal cells: H1 red, Hx blue, RH, yellow. Scale 20 μm.

ganglion cells are strongly E+, while the distinct set of displaced amacrine cells and occasional misplaced amacrine cells have very low E signals (Fig. 27A). The taurine signal of the ganglion cell layer is, in effect, the inverse of the glutamate signal with very low signals in ganglion cells and elevated signals in amacrine cells (Fig. 27B). The optic fiber layer presages the massive τ+ signature of the optic nerve proper (see below), due to optic nerve astrocytes and not the optic fibers themselves. The GABA signal selectively displays displaced and misplaced amacrine cells. The simultaneous segregation of these signals are best displayed as γGE $\rightarrow$ *rgb* and **cmy** mappings (Fig. 28). In these mappings it is clear that ganglion cells display minor

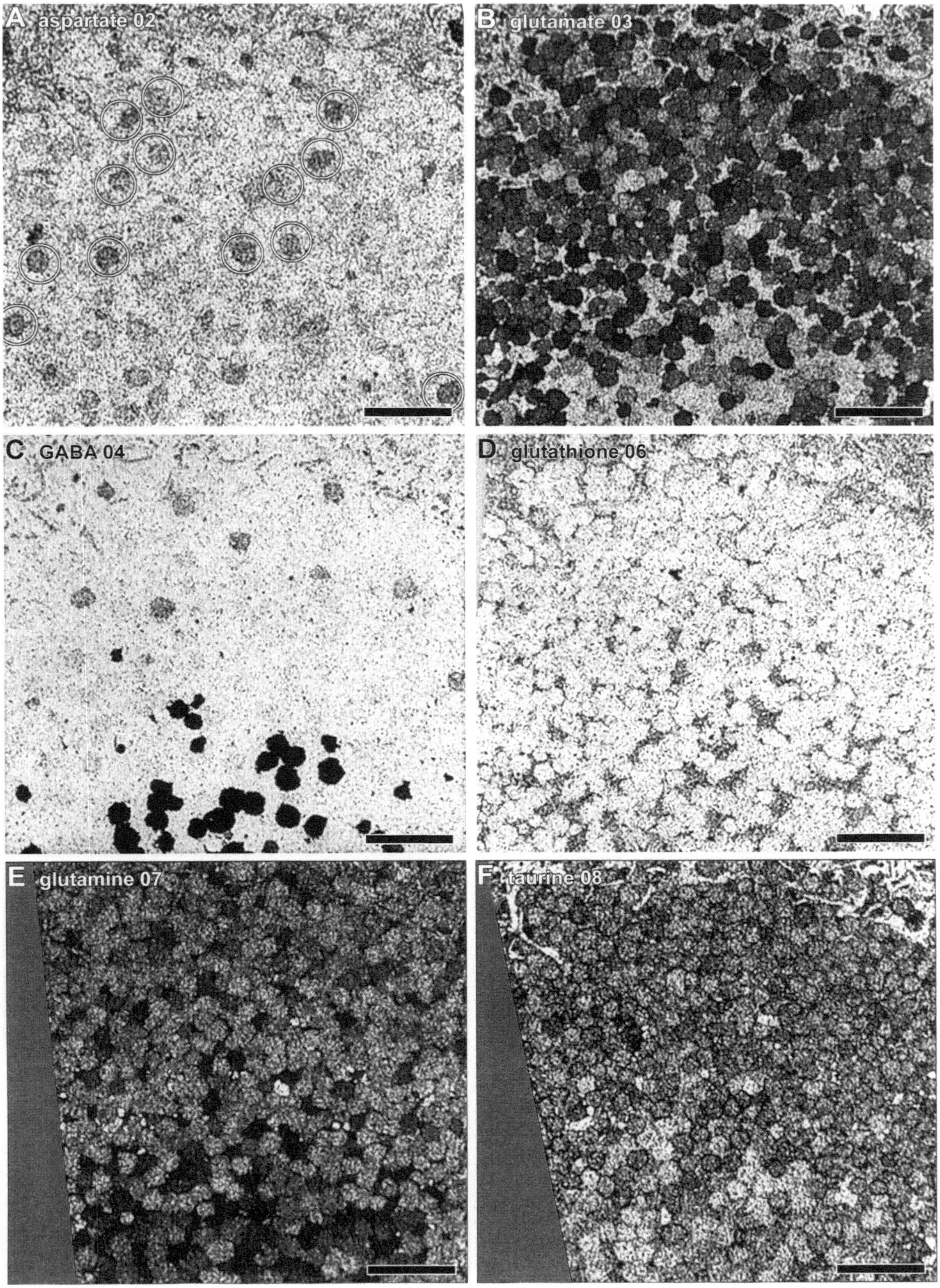

Fig. 19. A high-resolution library of the bipolar cell layer: DEJQ$\tau\gamma$ signals in the dark-adapted retina. A, aspartate 02. Aspartate-rich bipolar cells are circled. B, glutamate 03. C, GABA 04. Both strong and weak γ+ bipolar cell-like somas are evident. D, glutathione 06. E, glutamine 07. F, taurine 08. Section thickness: 250 nm. Scale 20 μm.

variations in their low taurine contents and that displaced and misplaced amacrine cells belong to different $\gamma+$ classes.

SUMMARY N-PLOTS

The arrays of registered monochrome and triplet mappings are the physical evidences of diverse molecular phenotypes between and within retinal cell types. Without such precision signal archives, decomposition of types cannot achieved. Even theme maps, such as in Figure 16, are difficult, if not impossible, to interpret by back-and-forth comparison with the source data. N-plots of signatures for each bipolar, horizontal, and ganglion cell class/superclass are summarized in Figure 29. Six separable classes of bipolar cells emerged from isodata clustering. BC1 is the most remarkable and is a $\gamma+$ BC patterned in an array with a conformity ratio of >3, arguing that it is a singular type comprising but 3% of the entire bipolar cell cohort. The GABA content is only about 1 mM, much lower than that of any amacrine cell but equivalent to H1 horizontal cells. And the glutamate content is submillimolar, 2–10× lower than any other bipolar cell, but equivalent to the contents of light-adapted cones. What is the likely neurotransmitter of this cell? Arguably, it could be GABA or glutamate. The remaining bipolar cells are separable based on their characteristic, complex mixtures of glutamate, aspartate, glutamine and taurine. The ON-center mixed rod-cone Bb bipolar cells are likely contained in classes BC5 (Bb1, Bb2) and BC6 (Bb3), each comprising about 12% of the entire bipolar cell cohort. The remaining bipolar cells are all contained in classes BC2,3,4. Superclass BC3, the lowest DQ composite signal group in the E+ bipolar cells, makes up 46% of all bipolar cells and likely contains most or all of the cone bipolar cells. It is also a collection of the smallest cells in the inner nuclear layer, which matches well with the cone bipolar cell profiles of other cyprinids. Using the class/superclass minimum fractions as a representative size for pure cone bipolar cells, it is clear that this diversity could harbor over 20 classes: easily accomodating the estimated 15 cell classes described for other cyprinids (Sherry & Yazulla, 1992).

The ganglion cell population is likely as complex as in the goldfish (Hitchcock & Easter, 1986), but similar to the goldfish, the signatures do not segregate them effectively into classes or superclasses. Ganglion cells display a remarkably stable and distinctive signature, with a high DQ composite and very low AJ and Gτ

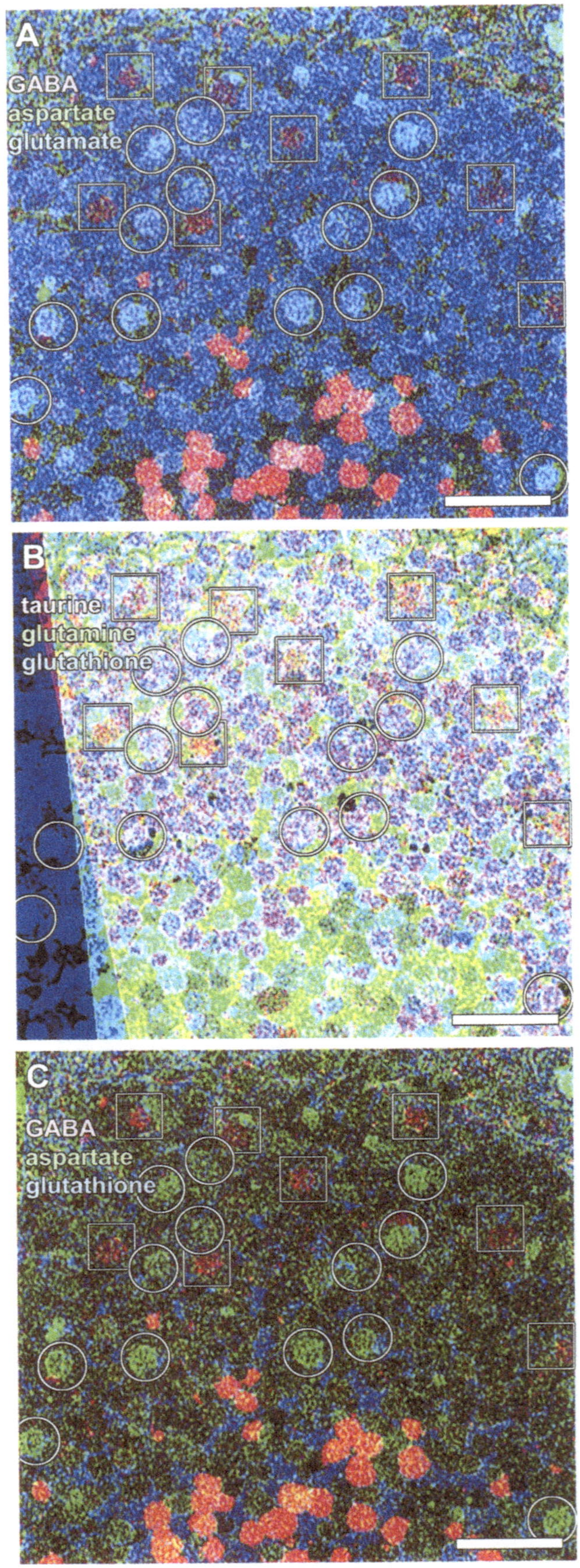

Fig. 20. Triplet mappings of the bipolar cell layer: A, γGE $\rightarrow$ *rgb*. Aspartate-rich cells are circled, GABA-rich cells are boxed. τQE $\rightarrow$ *rgb*. C, γDJ $\rightarrow$ *rgb*. Scale 20 μm.

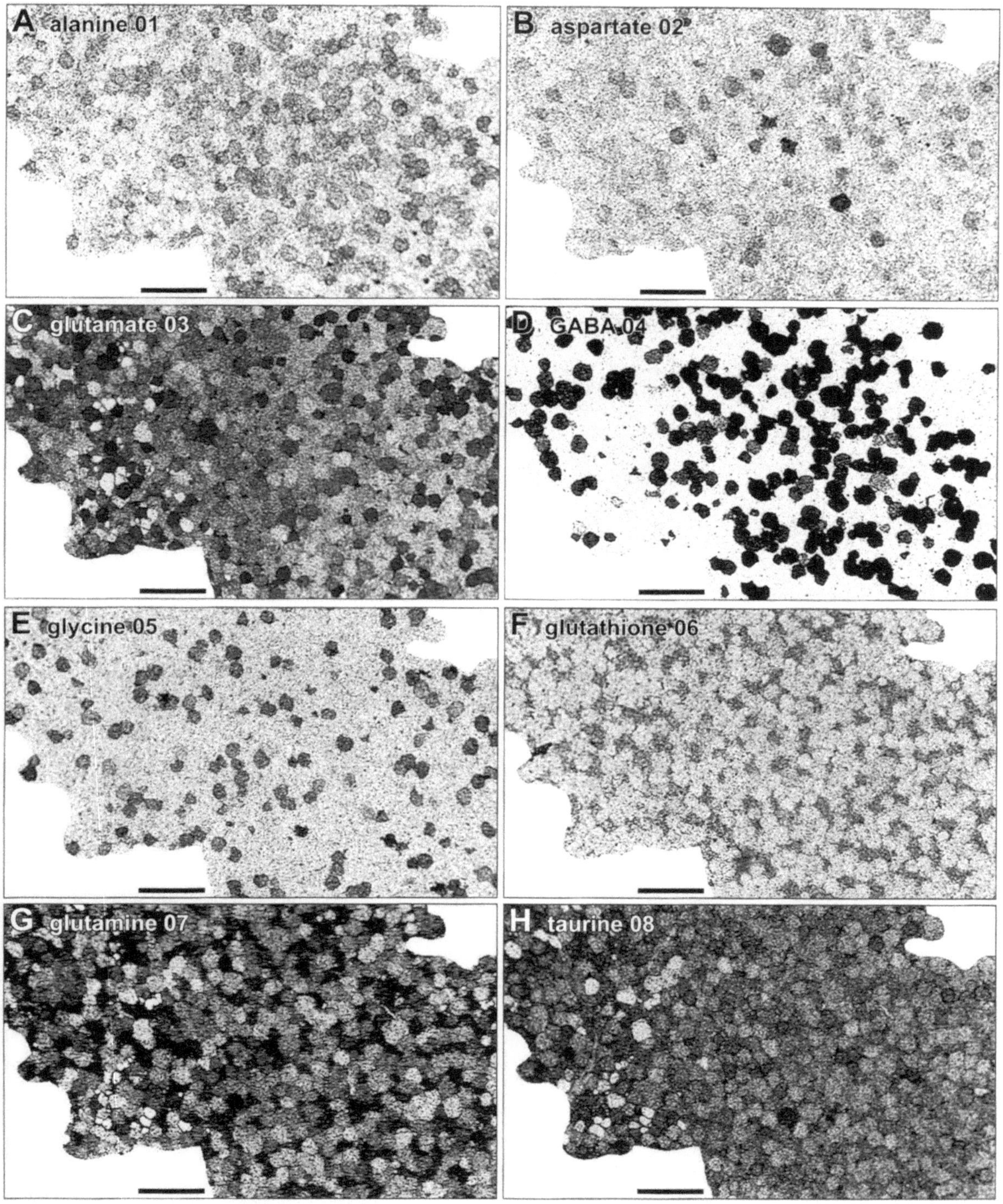

Fig. 21. A high-resolution library of the amacrine cell layer: ADEGJQ$\tau\gamma$ signals in the dark-adapted retina. A, alanine 01. B, aspartate 02. C, glutamate 03. D, GABA 04. E, glycine 05. F, glutathione 06. G, glutamine 07. H, taurine 08. Section thickness: 250 nm. Scale 20 μm.

composites. This signature is also characteristic of displaced ganglion cells in the goldfish (Marc *et al.*, 1990), and examples of similar cell signatures were detected in the zebrafish.

The neurochemistry of presumed lateral "inhibitory" horizontal cells remains one of the mysteries of retinal processing. There are three major horizontal cell signatures in the zebrafish retina. H1 horizontal cells are the γ+ cells of the horizontal cell layer, but might as well be termed E+, as they contain up to 1 mM glutamate, as in goldfish and other cyprinids. Based on signal content and the presence of H1 horizontal cell

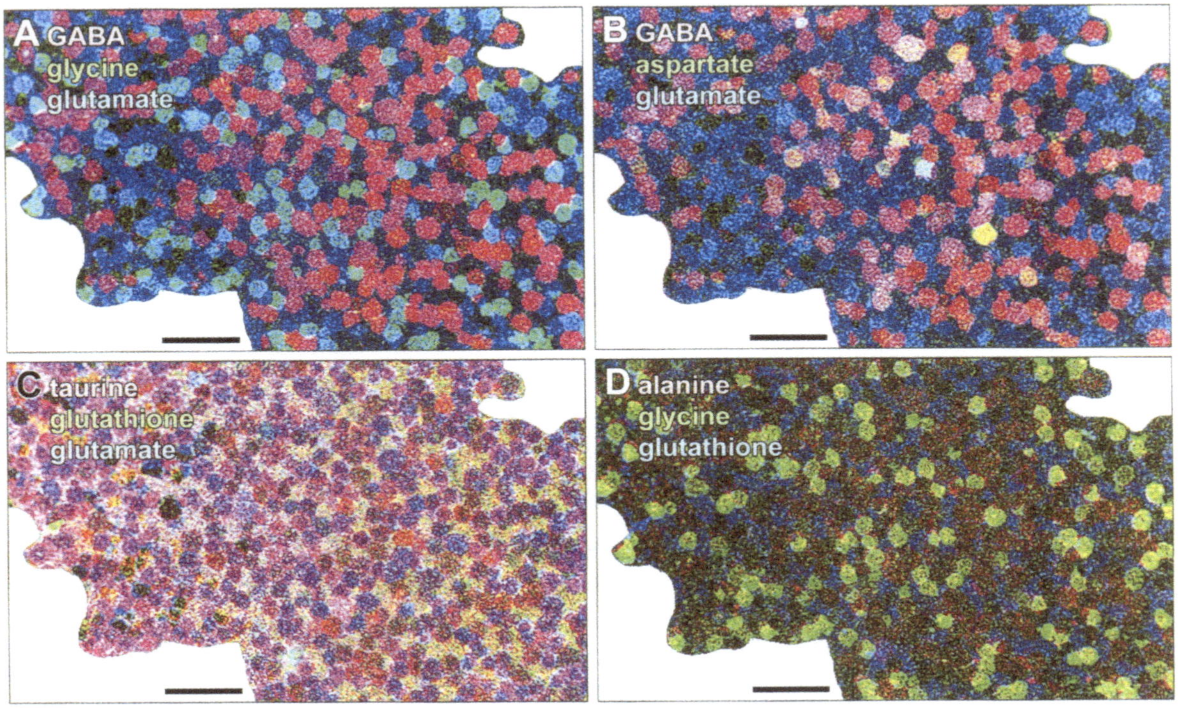

Fig. 22. Triplet mappings of the amacrine cell layer: A, γGE → *rgb*. B, γDE → *rgb*; C, τQE → *rgb*. D, AGJ → *rgb*. Scale 20 μm.

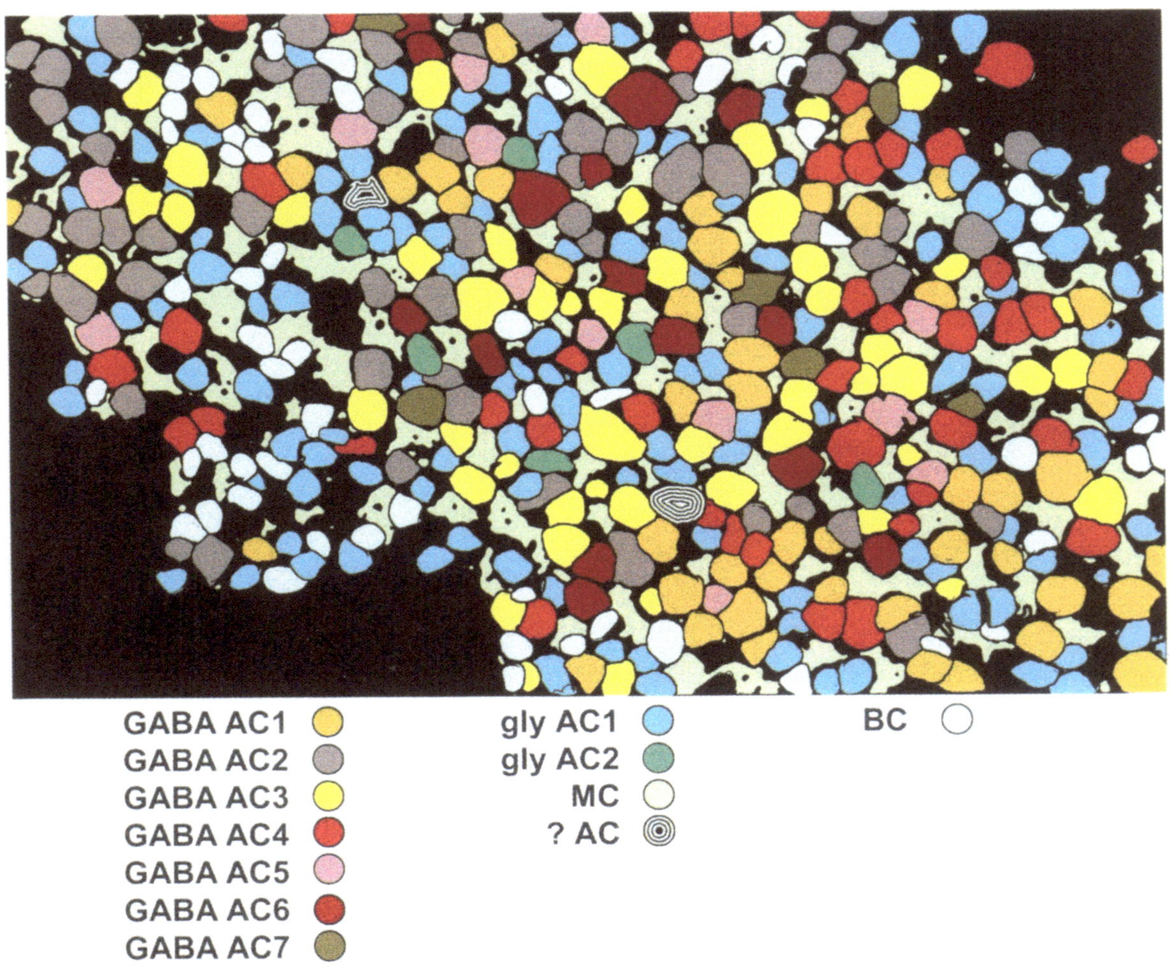

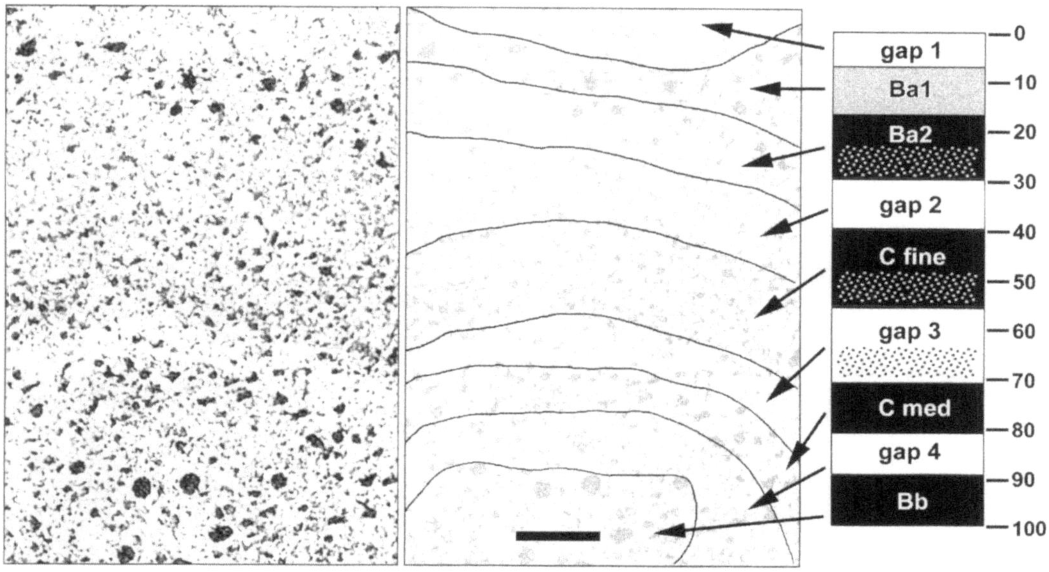

Fig. 24. Bipolar cell axon terminal arrays in the inner plexiform layer. Different morphological types of bipolar cells target distinct levels in the inner plexiform layer and, in combination with other signals, a formal lamination can be extracted. Left panel, the median-filtered γ-E difference signal in an oblique, high-resolution section isolates bipolar cell terminals. The center panel superimposes lamination borders on a brightness remapped image. The right panel is a formal lamination scheme where white *gap* boxes indicate a relative paucity of bipolar cell terminals. Dark boxes indicate bipolar cell rich zones: Ba1, class Ba1 OFF-center bipolar cells; Ba2, class Ba2 OFF-center bipolar cells; C fine, fine cone bipolar cell terminals; C med, coarser cone bipolar cell terminals; Bb, class Bb ON-center bipolar cell terminals. Stippling denotes finer scale subdivision of a region: levels 25 and 65 are zones where pairs starburst and pyriform radiate cholinergic amacrine cells arborize; level 50 is where a unique set of glycinergic amacrine cells terminate.

synaptic specializations in the outer plexiform layer targeting glycinergic interplexiform cell, H1 horizontal cells could as readily use glutamate as GABA, or both, as a signal molecule, in addition to pure biophysical signal modes (Kamermans & Spekreijse, 1999). The case for a specific molecule is far from established and becomes more complicated when the modes of signaling of Hx or RH cells are considered: both contain little GABA, but rod horizontal cells in particular contain about 3 mM aspartate and glutamate. So even though distinctive signatures are evident from the N-plots, the neurochemistry of horizontal cells in general remains puzzling. As noted earlier, the size of the Hx group is large enough to justify the speculation that more than one class of horizontal cell is submerged in the set.

There are seven groups of γ+ and two of G+ amacrine cells (Fig. 30). Classes γ AC1 and γ AC3 each comprise $\approx$30% of the γ+ amacrine cell cohort and are thus likely to be superclasses. Among several differentiating features, class γ AC1contains 5–10$\times$ less glutamate than class γ AC3. Class γ AC3 is also one of the low-taurine cells so evident in the triplet displays of the amacrine cell layer. The next largest classes are γ AC4 (14%) and γ AC6 (10%). Class γ AC4 is unique in a mathematical sense: no single signature sets it apart, but N-dimensional vector mapping segregates it from other classes. Class γ AC6 is the high-aspartate cohort of the amacrine cell layer. The smallest classes, γ AC2 (6%), γ AC5 (6%) and γ AC7 (4%) each have a unique signature, with γ AC5 being the lowest γ signal population in the γ+ cells of the amacrine cell layer. The glycine signals of G+ amacrine cells are modest compared to the primary signals of other neurons and they also contain significant glutamate signals. As noted above, the G AC1 group is certainly a superclass, as evidenced by multimodal DQ components of the N-plot. This illustrates that even univariate deconvolution has limits, as segregating the DQ cluster would require a bivariate deconvolution algorithm, which we have not yet implemented. Class G AC2 is a very small group (5% of the G+ cohort, only 1% of the amacrine cell type) with weak but significant glycine levels, $\approx$800 μM. We place

Fig. 23. A formal theme map of the dark-adapted amacrine cell layer. This map displays the dataset from Figures 22 and 23. Color-code as in Figure 16, with additional codes for unidentified cells. Bipolar cells are grouped as a single type and not classified for simplicity.

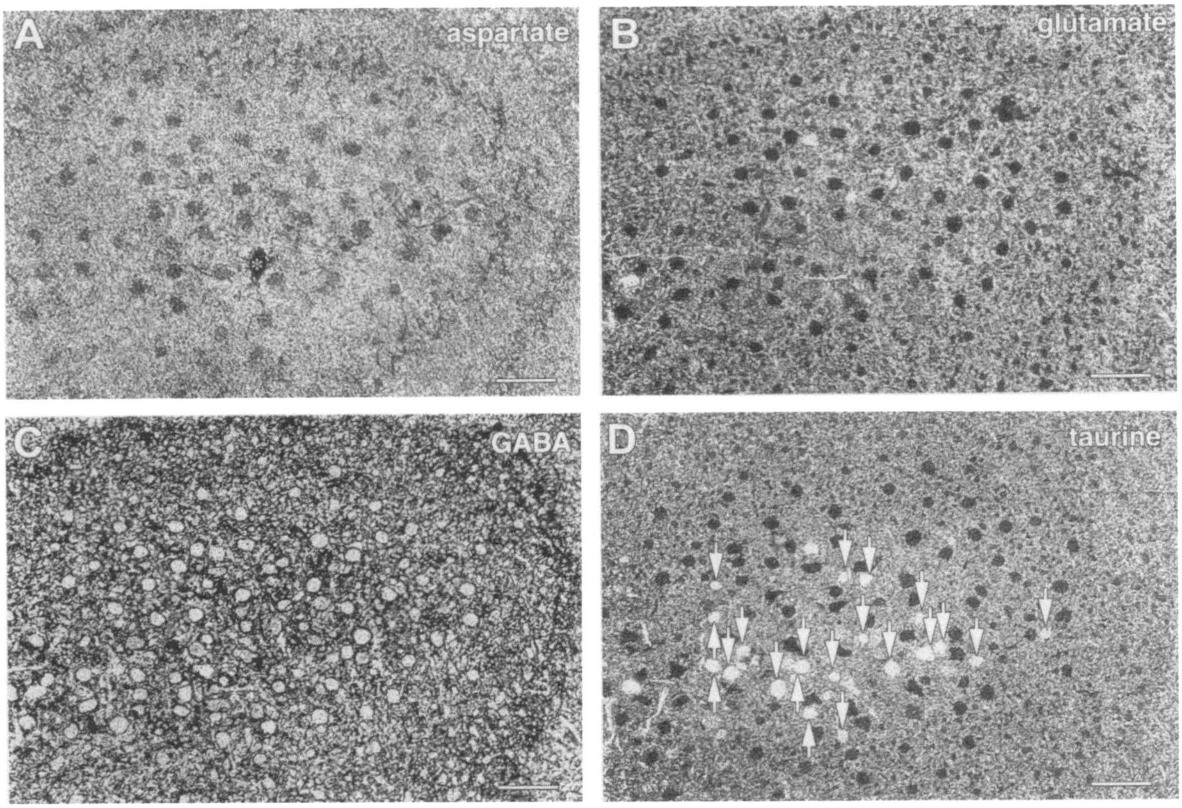

Fig. 25. A high-resolution library of the Bb bipolar cell terminal layer at level 90 of the inner plexiform layer: DE$\tau\gamma$ signals in the dark-adapted retina. A, aspartate. Asterisk denotes an image defect. B, glutamate. C, GABA. D, taurine. Arrows indicate ganglion cells. Section thickness: 250 nm. Scale 20 μm.

it in the G+ group for circumstantial reasons: (1) the GABA content is nil; (2) although it has significant glutamate content, so does many an amacrine cell; (3) it does not seem likely to be a bipolar cell based on its slightly lower τ and higher J signals; (4) it is not likely to be a dopaminergic cell, due to its high frequency and small soma size; (5) it is unlike any ganglion cell in its taurine content. We must admit, however, that we really do not know what it is. Even discounting these cells, γ+ amacrine cells comprise 63% and G+ amacrine cells 35% of the neurons in the amacrine cell layer proper. Thus any developmental, experimental or pathologic processes that differentially impact these populations can easily be tracked in terms of their effects on the proportions of γ+ and G+ cells.

SIGNALS IN THE LIGHT-ADAPTED NEURAL RETINA

On the whole, there is little *obvious* difference in the signals of light and dark-adapted retinas, but since there are real changes in the signatures of LD and SD cones, it was obviously necessary to screen the neural retina. Subtle patterns of signal alteration were detected with, we believe, interesting consequences for the physiological properties of amacrine cells in particular. Most basis signals viewed in single-channel mode for light-adapted retinas (Fig. 31) were indistinguishable from dark-adapted specimens (Fig. 10). Alanine (Fig. 31A), glutamate (Fig. 31C), GABA (Fig. 31D), glycine (Fig. 31E), glutamine (Fig. 31G) and taurine (Fig. 31H) signals display the same patterns and signal strengths appeared unchanged. In three of five samples, the glutathione levels (*e.g.* Fig. 31F) appear slightly higher in the inner plexiform layer, as though the overall Müller cell content was enhanced. The aspartate signals were dramatically different in all cases (Fig. 31B): all horizontal cell aspartate levels dropped, and the levels in ganglion cells and bipolar cells rose. The significance is unknown, of course. One major change is invisible in this mode, however, and required triplet mapping (Fig. 32) before it was found. In the light-adapted retina, there is strong dual G+ γ+ signal localization (Fig. 32A) in cells found to be purely G+ when dark-adapted (Fig. 11A). In fact, nearly all G+ amacrine cells showed some suprathreshold GABA signal, usually at the 300–500 μM level. However, one set of cells showed much stronger GABA signals ($\approx$1 mM) combined with 3–5 mM glycine (Fig. 32A). These may be the homologues of the Gγ amacrine cells in the goldfish retina (Marc *et al.*, 1995). There are even

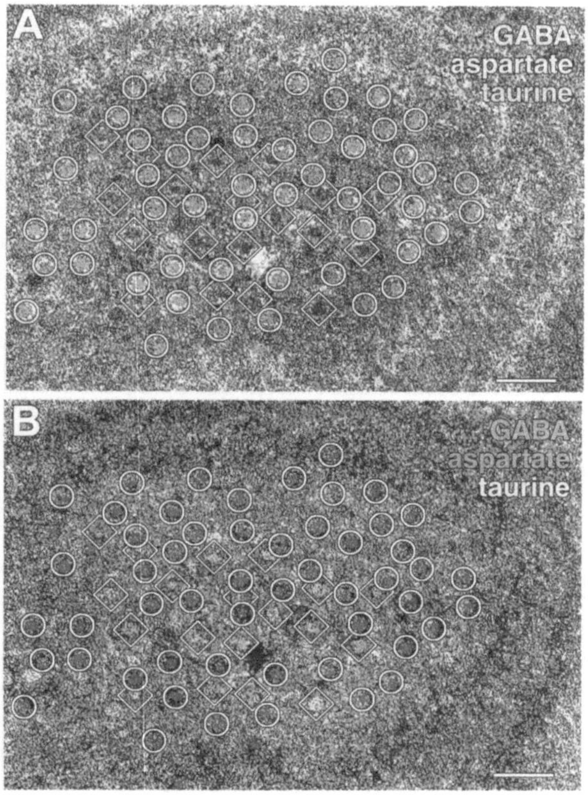

Fig. 26. Triplet mappings of the Bb bipolar cell terminal layer: A, $\gamma D\tau \rightarrow$ *rgb*. Circles denote Bb1 and Bb2 bipolar cell terminals. Diamonds denote Bb3 bipolar cell terminals. B, $\gamma D\tau \rightarrow$ **cmy**. Scale 20 μm.

some very strong γ+ amacrine cells and E+ bipolar cells that show enhanced glycine signals, and the latter have also been detected in the goldfish retina. These are subtle phenomena, not obvious without detailed signature analysis and N-plots, and their significances will be addressed in the discussion. Even with this change in signature, G+ amacrine cells remain strongly A+ (Fig. 32B, AGJ $\rightarrow$ *rgb*), continuing the theme of alanine/glycine association. Moreover, the light-adapted τQE $\rightarrow$ *rgb* mapping is indistinguishable from the dark-adapted retina, supporting the idea metabolic changes are selective, not global. The inclusion of the aspartate channel (Fig. 32C, γDE $\rightarrow$ *rgb*) does produce a different and unique image, lacking the distinctive horizontal cell band and with more pronounced bipolar cell and ganglion cell definition. The role of elevated aspartate levels in any cells, much less neurons, is without explanation.

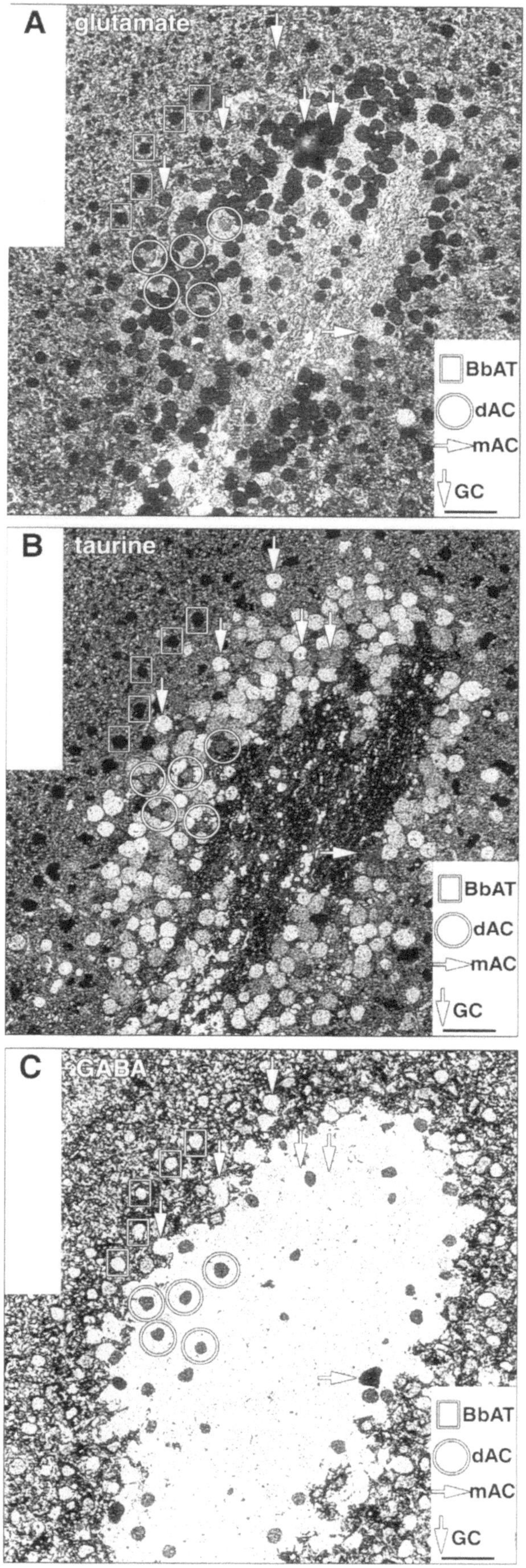

Fig. 27. A high-resolution library of the ganglion cell layer: $E\tau\gamma$ signals in the dark-adapted retina. A, glutamate. B, taurine. C, GABA. Abbreviations: dAC, displaced amacrine cell; mAC, misplaced amacrine cell. Section thickness: 250 nm. Scale 20 μm.

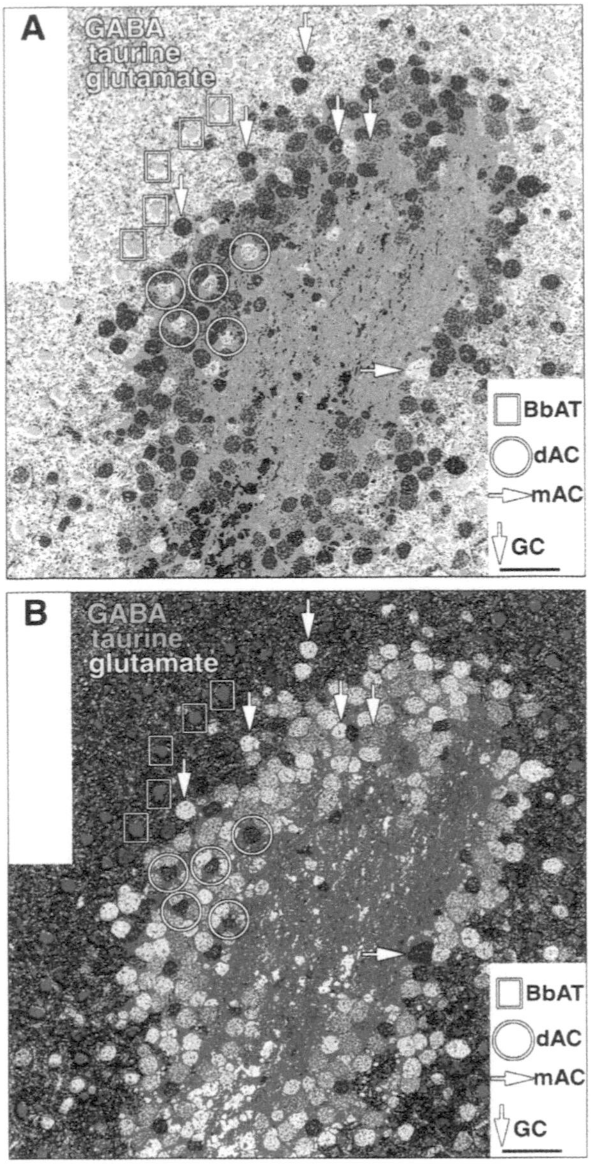

Fig. 28. Triplet mappings of the ganglion cell layer: A, $\gamma\tau E \rightarrow$ *rgb*. B, $\gamma\tau E \rightarrow$ **cmy**. Scale 20 μm.

SPECIALIZED AREAS OF THE RETINA: THE OPTIC NERVE HEAD AND OPTIC NERVE

The ganglion cell axon outflow into the axon fascicles of the peripheral retina, merging into the optic fiber layer, the optic nerve head and the heavily fasciculated optic nerve (Easter *et al.*, 1981), a rough folded ribbon, also have distinctive phenotypes. Figure 33 displays a registered data set from a dark-adapted zebrafish retina encompassing all of these structures, the vasculature of the optic nerve proper and the choroidal rete. Even relatively weak patterns such as alanine and aspartate signals (Fig. 33A and B) illustrate the delicate changes in molecular signatures of the optic fiber layer, head and nerve change; various signals are graded accross the path of the folded ribbon. The optic nerves of many fishes resemble flat ribbon cables crumpled into compact bundles. The thick, most extended portion is made of axons from the central retina which is the oldest part of the eye: this organization is largely chronotopic, not precisely retinotopic. The thinnest, most folded segment ends in axons from the peripheral retina. The glutamate image (Fig. 33C) validates the expectation that the axons of ganglion cells ought to contain high glutamate signals, but also shows a surprising glutamate-rich cap at the peripheral end of the ribbon, which may be the pale nonmyelinated fiber tract (the newest axons) in goldfish (Easter *et al.*, 1981) or a zone nearby. Vascular endothelia of vitreal vessels, optic nerve vessels, the choroidal rete, the choroids and choriocapillaris all have glutamate levels as high as any retinal neuron. Conversely, high glycine signals seem to be the signature of *bona fide* glycinergic neurons, and no element in the optic nerve or its associated tissues has any significant glycine content (*i.e.* all levels must be below 50 μM). Glutathione signals (Fig. 33E) display some intriguing specializations. The glutamate-rich cap at the peripheral limit of the optic nerve ribbon is extremely J+, second only to the RPE, which suggests that this cap contains glial cells. There is also a subtle increase in glutathione levels within the body of the ribbon, following the same gradient as alanine and aspartate. Finally, the glial (or perhaps ependymal) derivatives that form the margin of the optic nerve head in the retina also display significant elevation in glutathione. GABA levels in the optic nerve and associated tissues are nil, except for a subtle elevation in the cap (Fig. 33F). Glutamine levels are massive throughout the optic nerve and in the cap, consistent with both axonal and glial sources (Fig. 33G). It is again worth noting that endothelial glutamine levels are quite low (as are aspartate levels), perhaps suggesting that these metabolites are not the prime sources of endothelial glutamate. Finally, taurine signals create a rich tableau (Fig. 33H). Taurine levels in the optic fiber layer rival those of the outer retina, and are likely representative of intraretinal fiber layer astrocytes. But as the axons flow into the optic nerve head, taurine signals weaken by nearly 100-fold, suggesting that glia (astrocytes) with an entirely different metabolic program now support the axonal extracellular regime. The optic nerve is reticulated with τ+ elements (especially near the surface of the nerve), but no central $\rightarrow$ peripheral signal gradient is evident. Again, the unmyelinated cap has extremely high taurine signals, approaching 10 mM or higher. Triplet mapping uncovers a wealth of signatures unmatched by any other visualization process or probe set. The standard γGE $\rightarrow$ *rgb* mapping is unremarkable, but for the enhancement of the subtle signals in the peripheral optic nerve glial cap (Fig. 34A). In contrast, the τQE $\rightarrow$ *rgb* and τQJ $\rightarrow$ *rgb* triplets (Fig. 34B and C) reveals a rich polychromatic display of signatures. Only one area will be noted here: the fine discrimination

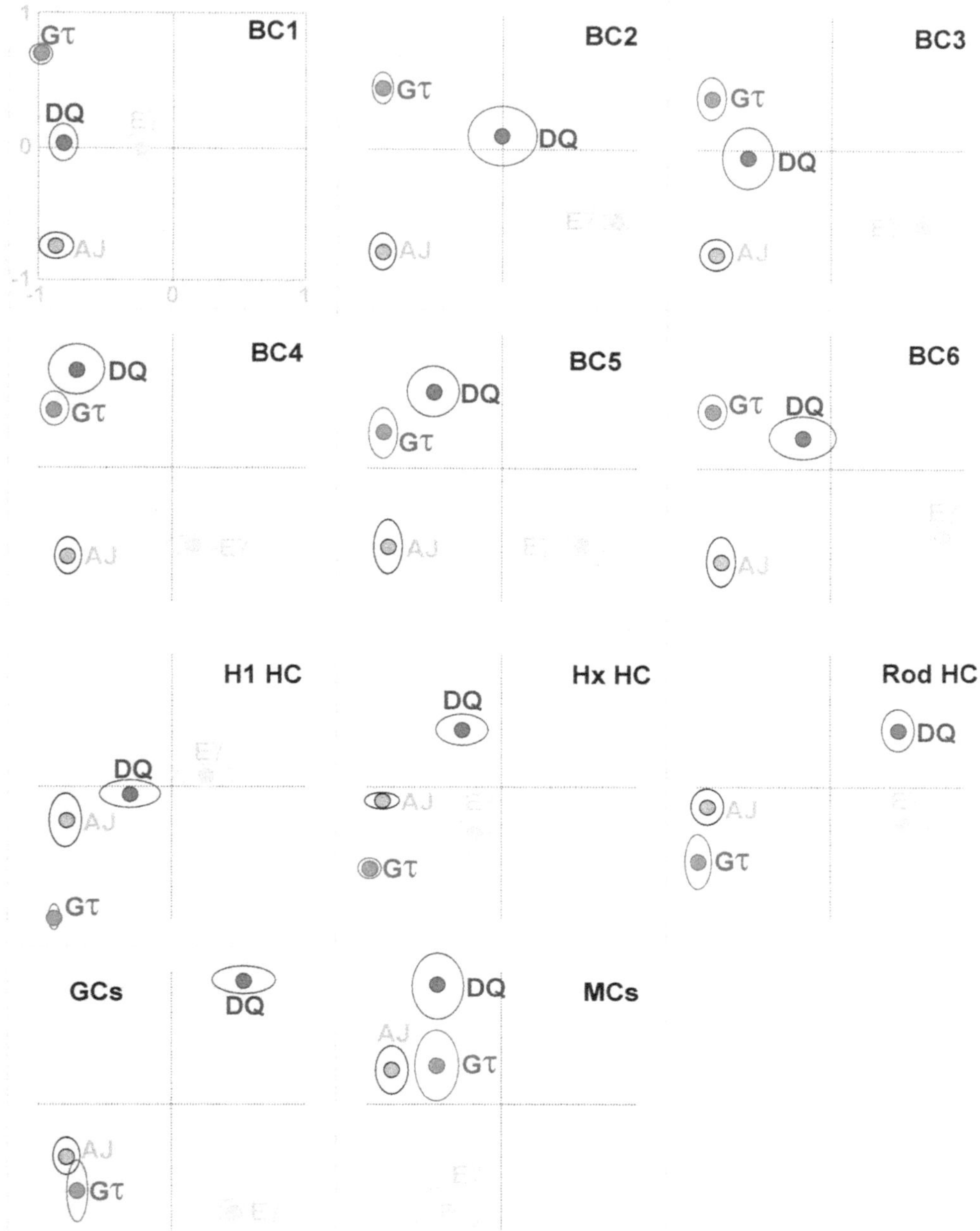

Fig. 29. Horizontal, bipolar, ganglion and Müller cell molecular phenotyping N-plots. Each plot is coded as in Figure 2 and is derived from dark-adapted retina.

of molecular phenotype signals of the axonal core, the glial sheath and terminal glial cap of the peripheral optic nerve. The cap displays a unique mixture of maximal glutamate, taurine, glutamine and glutathione signals. No neural or glial element in the retina and no epithelial or endothelial cell has a similar signature. These strong signals obscure the fine gradient of minor metabolites, which can be easily expressed with a $\gamma D\tau \rightarrow rgb$ mapping (Fig. 34D). As various insets display (Fig. 34A, C and D), the retinal signatures remain precise across specimens, preserving characteristic neuronal and epithelial signals.

SPECIALIZED AREAS OF THE RETINA: THE MARGINAL ZONE

Of all regions, the circumferential marginal zone (also known as the germinal zone) of cyprinid retinas is perhaps the most provocative, as it contains adult retinal stem cells from which new photoreceptors and neurons

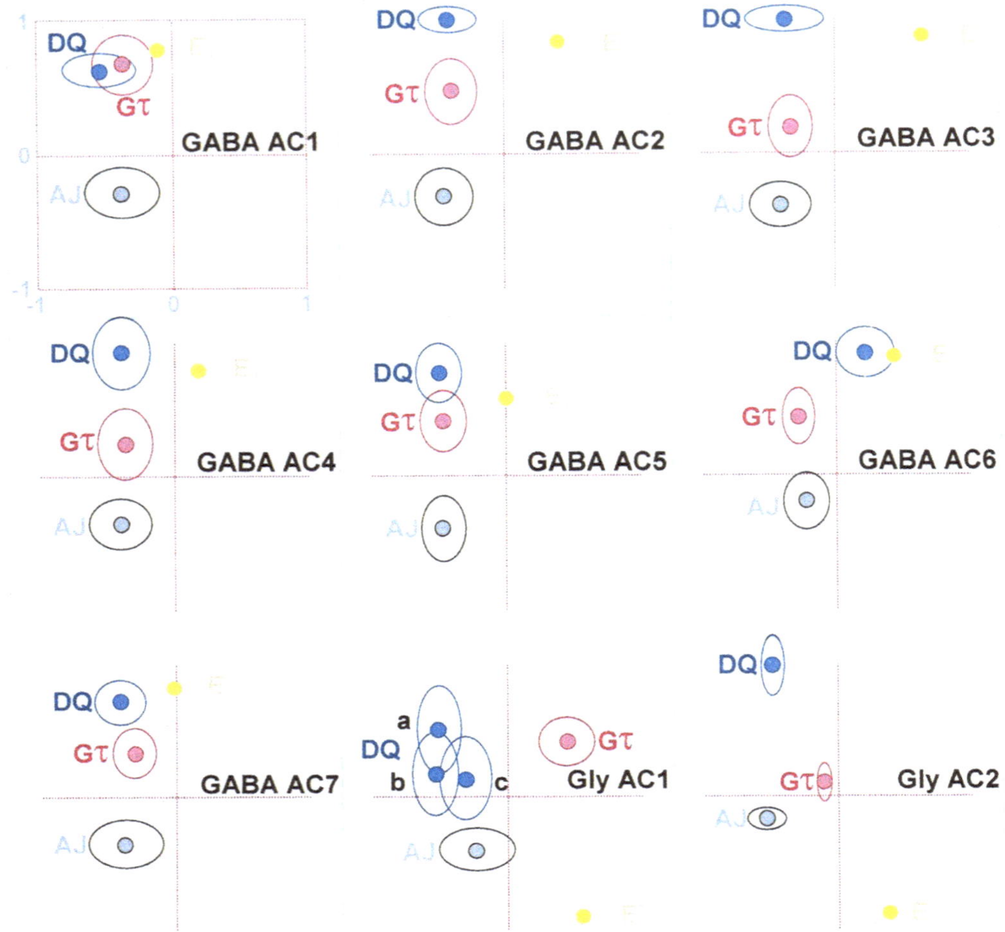

Fig. 30. Amacrine cell molecular phenotyping N-plots for 7 GABA and 2 glycine amacrine cell classes. Each plot is coded as in Figure 2 and is derived from dark-adapted retina.

are generated. We performed molecular phenotyping with the same approaches as described for all other retinal loci and the optic nerve. The entire region can be visualized as the marginal retina abutting the marginal zone and the continuous transition from retina to the epithelial ciliary zone and iris (Fig. 35: A, alanine; B aspartate; C, glutamate; D, glycine; E, GABA; F, glutathione; G, glutamine; H, taurine). The interface between the neural retina proper and the formal marginal zone is abrupt, with no detectable gradients on first examination. The processes of amacrine cells (Fig. 35D and E), bipolar and ganglion cells (Fig. 35C) do not invade the marginal zone. The retina itself appears largely mature right up to the marginal zone, with some exceptions. Aspartate signals (Fig. 35B) in the outer plexiform layer seem not to be stable for all horizontal cells until 75–100 μM away from the marginal zone and the precise G+ amacrine cell lamina at level 50 of the inner plexiform layer seems absent until 50 μM from the marginal zone. Cells likely to be ganglion cells based on their EQ signatures (Fig. 35C and G) are present in the ganglion cell layer right up to the border and within the proximal part of the marginal zone. The marginal zone itself is an apparently continuous flattened torus, about 50–60 cells in cross-section: only a fraction are likely neuronal precursors. The zone is particularly well-delineated by its moderate glutamine and taurine signals. Proceeding further anteriorly, the marginal zone gives way to the ciliary body. In the zebrafish this is an uncomplicated area comprised of the distal PCE and proximal NPCE bilayer. The signals of these epithelial sheets generally resemble the RPE, but some clear differences emerge, especially in glutathione, glutamine and taurine content. As the RPE approaches the peripheral retina its glutathione signals get slightly weaker, though they are still far elevated above any retinal cell. This signal is contiguous with the pigmented ciliary epithelium and the pigmented epithelium abutting the distal surface of the marginal zone. However, the NPCE emerges at the anterior edge of the marginal zone and

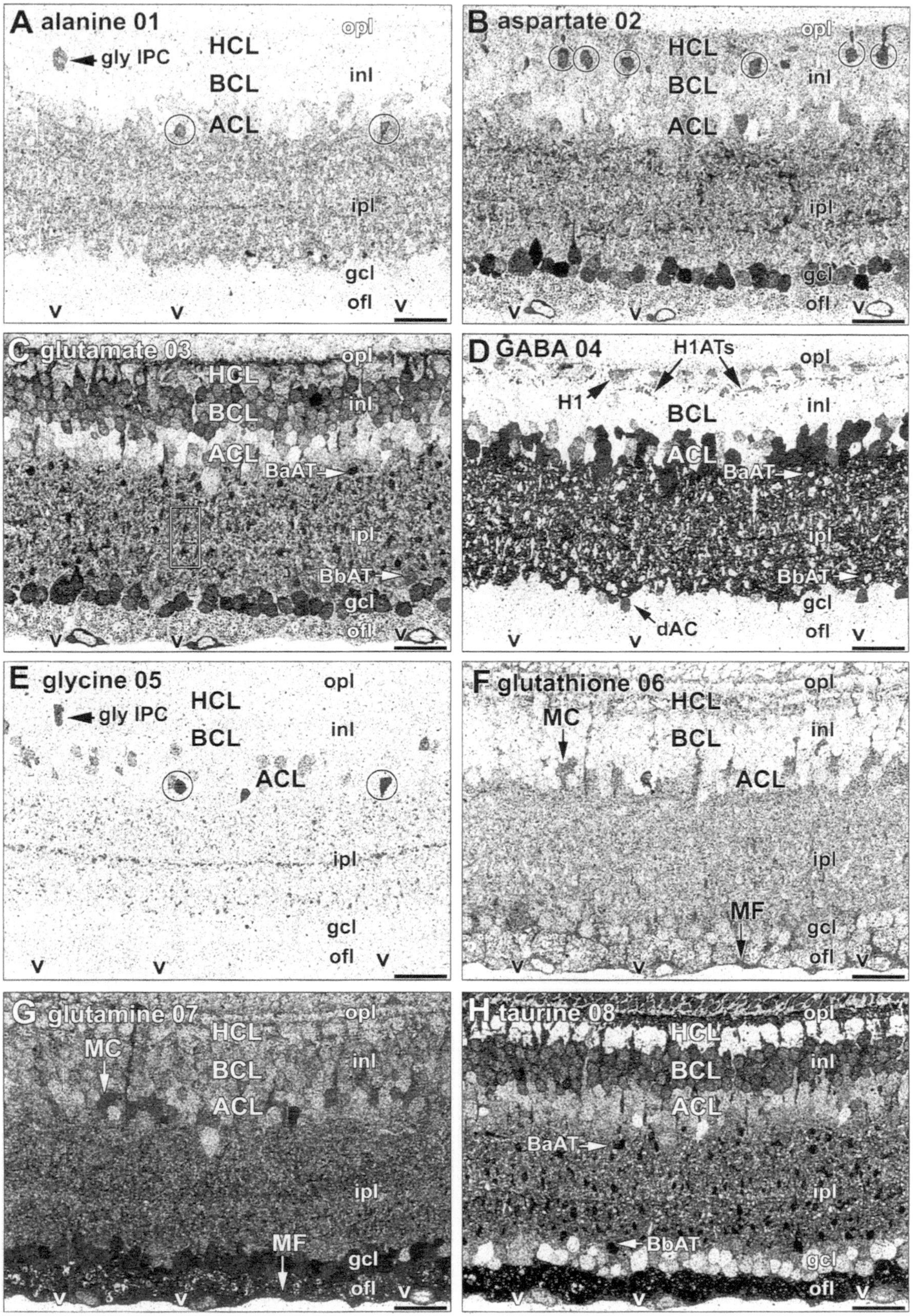

Fig. 31. Single channel, registered ADEGJQ$\tau\gamma$ signals in the light-adapted inner retina. A, alanine 01. B, aspartate 02. C, glutamate 03. D, glycine 04. E, glutathione 05. F, GABA 06. G, glutamine 07. H, taurine 08. Section thickness: 250 nm. Scale 20 μm.

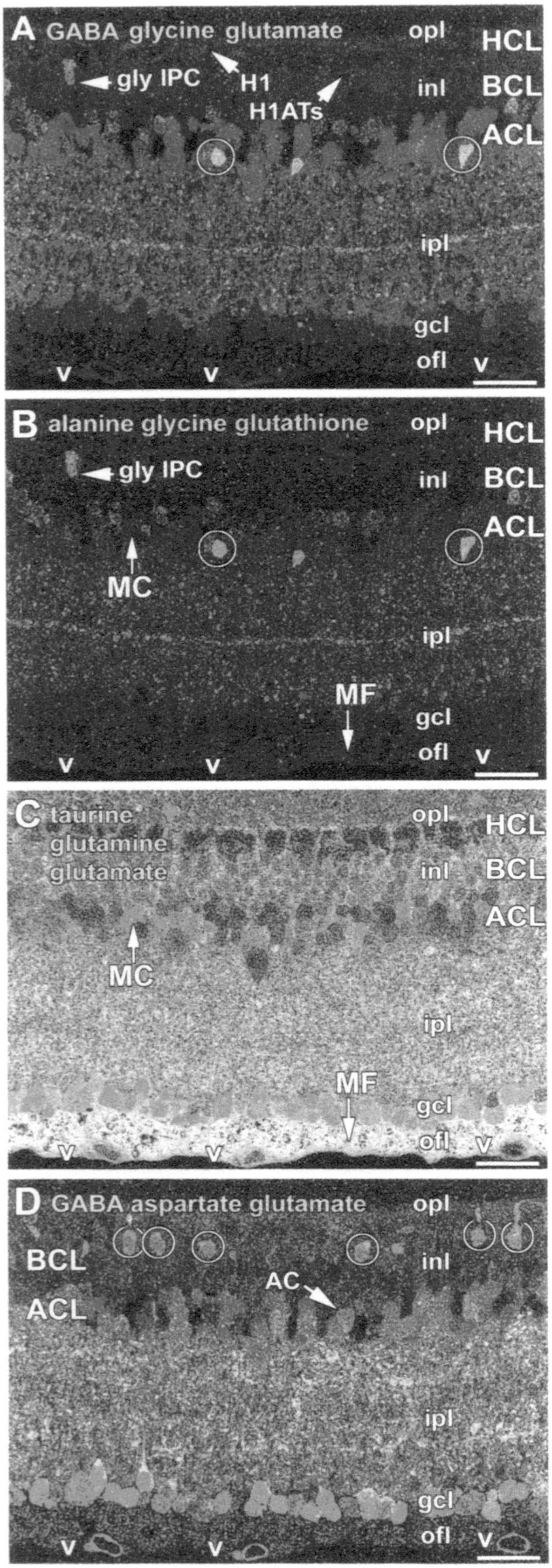

Fig. 32. Triplet mappings of the light-adapted inner retina. A, γGE $\rightarrow$ *rgb*. B, τQE $\rightarrow$ *rgb*. Scale 20 μm.

clearly has lower glutathione, higher glutamine and higher taurine signals (Fig. 35F, G and H). Since the PCE and NPCE are supposedly coupled by gap junctions at their interface (both apical surfaces point toward each other, unlike conventional transport monolayer epithelia), this signature difference must reflect either modulation of coupling or steady-state gradient differentials. The transition between the ciliary body and the iris epithelium becomes particularly evident in the glutamine signals (Fig. 35G). The choroid and the reflective tapetal layers of the eye have their own unique signatures, the most dramatic signal being the high glutamate enrichment (Fig. 35C). Taken together these signals allow the formation of a tissue-scale theme map (Fig. 36A) delimiting the cellular compartments of this complex area. The neural retina appear to possess its complete normal set of signatures (Fig. 36B, γGE $\rightarrow$ *rgb*; 36C, τQE $\rightarrow$ *rgb*), including the emergence of presumptive rod and cone horizontal cells at the retina:marginal zone interface (Fig. 36D, ADJ $\rightarrow$ *rgb*; 36E, τQD $\rightarrow$ *rgb*; 36F, γDτ $\rightarrow$ *rgb*). We define four sub-regions in and four epithelial types adjacent and anterior to the marginal zone.

Closer examination reveals substructure of the marginal zone and the young neurons apparently emerging from it (Fig. 37). In particular, the glutamate signals of the retina appear to fuse with or enter the distal half of the marginal zone, with diverse signal strengths evident within one cell of the marginal epithelium (Fig. 37C). The signatures of amacrine and ganglion cells (Fig. 38A, γGE $\rightarrow$ *rgb*) provide evidence that both cell populations develop distinctive, immature signatures within the marginal zone before being displaced into the neural retina proper and developing arbors. And perhaps more importantly, laminar position is already specified in the marginal zone. No cells with inappropriate laminar signatures appear in the the normal peripheral retina, even at the border with the marginal zone. Glutamate signals, combined with the glutathione (Fig. 37F), glutamine (Fig. 37G) and taurine signals (Fig. 37H), as well as τQE $\rightarrow$ *rgb* mapping (Fig. 38B) also suggest that most of the distal marginal zone cells apposed to the marginal epithelium are destined to be photoreceptors. The horizontal cell layer inserts deeply into the marginal zone where new rod horizontal cells also develop next to the photoreceptor precursors (Fig. 38C). The early maturation of ganglion cell-like signatures is particularly evident. In the proximal marginal zone, cells with strong glutamate signals (Fig. 37), but inappropriately low glutamine and high taurine and glutathione signals abut the true ganglion cell layer: these are likely immature ganglion cells. Within the ganglion cell layer, a few unequivocal ganglion cells show the same inappropriately high glutathione signals, but are also now developing the high glutamine contents of normal ganglion cells. The most peripheral of these still display inappropriate taurine signals, but neighboring cells lack taurine. We term these *maturing* ganglion cells. It is also quite obvious that different morphological types of ganglion cells emerge within 50 μm of the marginal zone. Taken together with classification, a detailed map of the marginal zone (Fig. 39) suggests that it is completely stratified into at least five distinguishable cohorts. Marginal zone class 1 cells (MZ1Cs) are those abutting the marginal epithelium (with their distal faces) and lined up next to the photoreceptor layer. They appear to be the transitional cell type already specified as photoreceptors. Next to those cells are class 2a (MZ2aCs) cells from marginal zone 2: cells both abutting the marginal epithelium and forming the core of the marginal zone. Class 2b cells are sandwiched between class 2a cells and the retina, appearing to be zone within which near-mature inner nuclear layer molecular phenotypes of horizontal, bipolar, γ+ amacrine, and G+ amacrine cells are activated. A thin class 2a strand only 1–2 cells wide emerges from the proximal cluster of 2a cells and is contiguous with the emerging ganglion cell layer. The proximal cluster of marginal zone cells in contact with the marginal epithelium, class 3 cells (MZ3Cs), evinces molecular phenotypes subtly different from the other zones and more like the marginal and ciliary epithelia. Thus we suspect it is the precursor pool for those transport layers. Finally, a complex of cells emerges from the vitreal surface of the marginal zone and protrudes slightly past the retina into the vitreous and in front of the NPCE. It appears sealed from the rest of the marginal zone by an interposed block of tissue with Müller cell signatures. Though we refer to it as marginal zone 4 and its associated class 4 cells, it may have nothing to do with the marginal zone. The signatures in MZ4 are extremely complex and contain nondescript cells with signal patterns very similar to vascular cells (Fig. 38). In sum, we find the marginal zone pre-laminated and detect no evidence of radial migration after emergence of maturing cell types into the retina proper. Additionally, arborization does not ensue until the cells enter the retina and even then new arbors never spread back into the marginal zone.

Discussion

Some 24 neural and 19 non-neural molecular phenotypes can be detected in the zebrafish retina and associated tissues (summarized in Tables 1 and 2). The zebrafish is, in most respects, a classic cyprinid. The only major differences between the zebrafish and its carp relatives are: (1) the better mosaic organization of the retina, with no LS cones filling in spaces between mosaic elements; (2) more distinctive laminar markers in the inner plexiform layer, especially the G+ band at level 50; (3) more diverse cellular signatures. The cells of the adult zebrafish retina are unfortunately about 20–50% smaller than those of other adult cyprinids, rendering physiological and some structural analyses more difficult. Nevertheless, clear homologies with the various neuronal and glial cohorts of other cyprinids are easily made.

PHOTORECEPTOR COHORTS AND MOLECULAR PHENOTYPES

The zebrafish ppears to lack the red LS, green LS and red miniature LS cones of goldfish, rudd, roach, carp and koi. The double cone mosaic is more compact and orderly than other cyprinids, with little room for these additional cell types. This is consistent with comparisons of mosaic structure in other teleost families: bottom grazers and predominant vegetarians (*e.g.* carp) have loose mosaics with additional "fill-in" cones and clear water predators have crystalline mosaics with exceptional conformity ratios. The blue SS cones of zebrafish have been termed LS cones by previous workers and this was reasonable in the sense that they are the largest of the "single" cones. However, the blue cones of the zebrafish are not homologous to LS cones but rather to the SS cones of other cyprinids; they match in shape, mosaic position, general pigment content and molecular phenotype. Similarly UV cones of the zebrafish, elsewhere termed SS cones, are the homologues of cyprinid UV-sensitive MSS cones in general, and not SS cones. The mosaic is unlike the loose square mosaic of the goldfish and is a row mosaic with precise interdigitation of double cones due to the 1/2 cycle offset of neighboring rows.

It appears that cyprinids share the same pattern of photoreceptor signatures and the same, somewhat unusual adaptation dynamics. Light-adapted rods and cones maintain very low levels of intracellular glutamate which increase dramatically (3–6$\times$) in rods and double cones, along with a decrease in both glutamine and taurine, upon dark-adaptation. The fact that both rods and cones respond in the same direction suggests that this metabolic change may have less to do with synaptic transmission than with some other regulatory process, such as changes in metabolic demands during sleep or in osmoregulation during the photomechanical responses. Since all photoreceptor signaling is glutamatergic, it is evident that cone synapses must be able to function and reload synaptic vesicles with a steady-state glutamate level of about 300 μM.

Connaughton *et al.* (1999) reported high glutamate levels in cones, apparently equivalent to bipolar and ganglion cells, but this was due to superposition saturation.

Both cones and rods contain high taurine signals. It is generally thought that the role of taurine, along with betaine and myo-inositol, is to act as part of a large pool of channel-permeant and transportable molecules for osmoregulatory adjustment in virtually all living systems (Moeckel *et al.*, 1997; Roberts, 2000; Schaffer *et al.*, 2000) and particularly in the retina (Morimura *et al.*, 1997). Taurine is the most abundant photoreceptor amino acid in light-adapted photoreceptors, ranging from 3–10 mM. Dark-adapted cones display up to a three-fold loss of taurine. The concomitant loss of glutamine is more explicable in terms of partial conversion to glutamate, though it is not simply stoichiometric: in LD cones, the actual drop in glutamine is about 500 μM while the increase in glutamate is some 4 mM. It is important to think of amino acid pools as open networks and the signals captured in any one state as reflective of steady states far from equilibrium. Simple additivity is unlikely to explain changes in most metabolic signatures. The general metabolic signature of zebrafish photoreceptors is similar to that of most vertebrates: a modest glutamate content and high taurine levels. The major differences with tetrapod and mammalian photoreceptors in particular are that the latter generally maintain low millimolar rather than micromolar glutamate, and show no adaptation dynamics.

HORIZONTAL CELL COHORTS AND MOLECULAR PHENOTYPES

Molecular phenotyping is particularly useful with inner retinal neurons, as it is one of the most incisive and general of tools for establishing homologies. The horizontal cells of the zebrafish retina are extremely small, with soma diameters of about 10 μm and population densities 2–2.5× those of other cyprinids. But it is clear that at least three distinctive populations are present with signatures similar to other cyprinids and non-mammalians. H1 horizontal cells are weakly γ+ E+ and τ−, as are the presumed long-wave cone

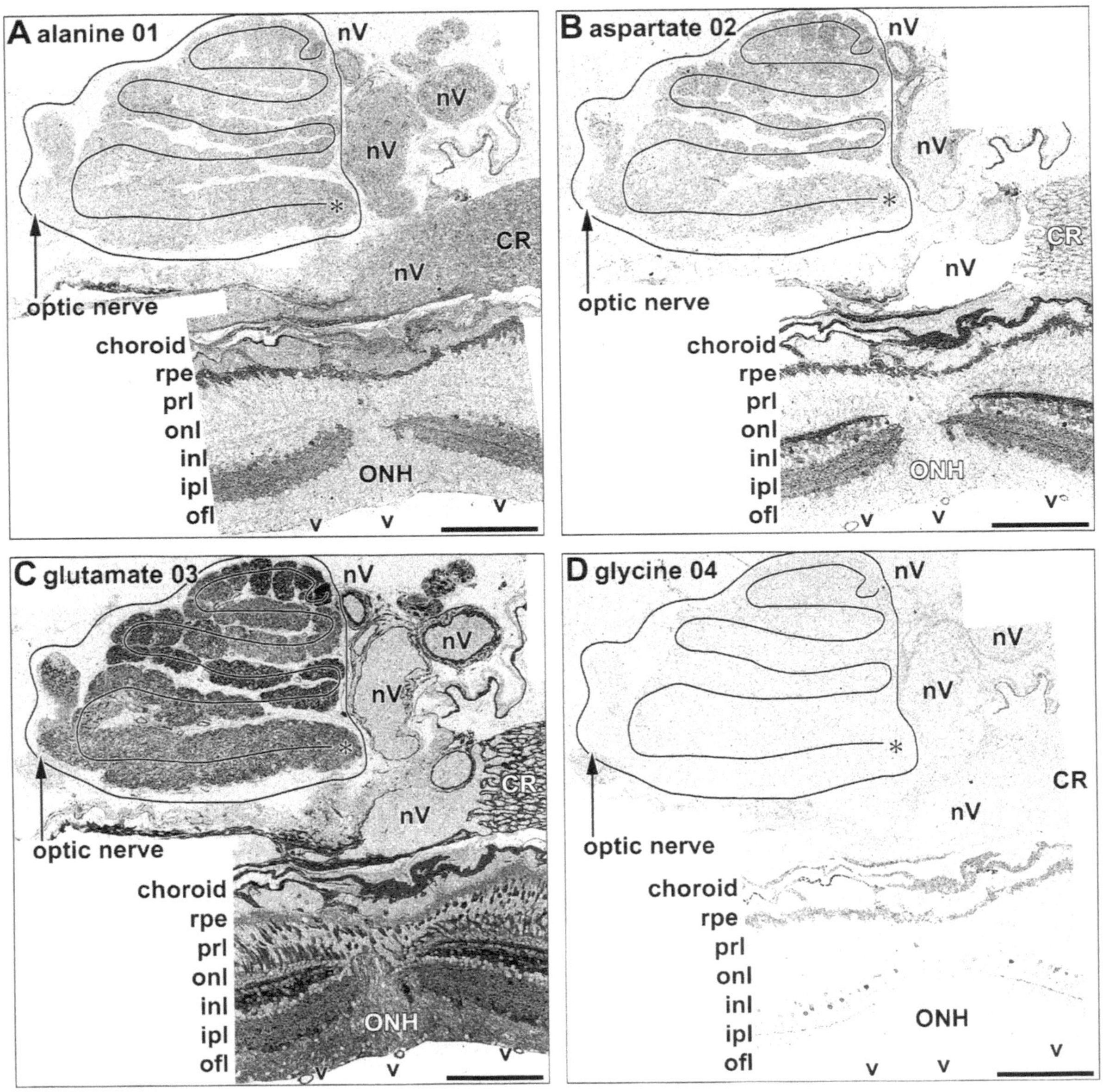

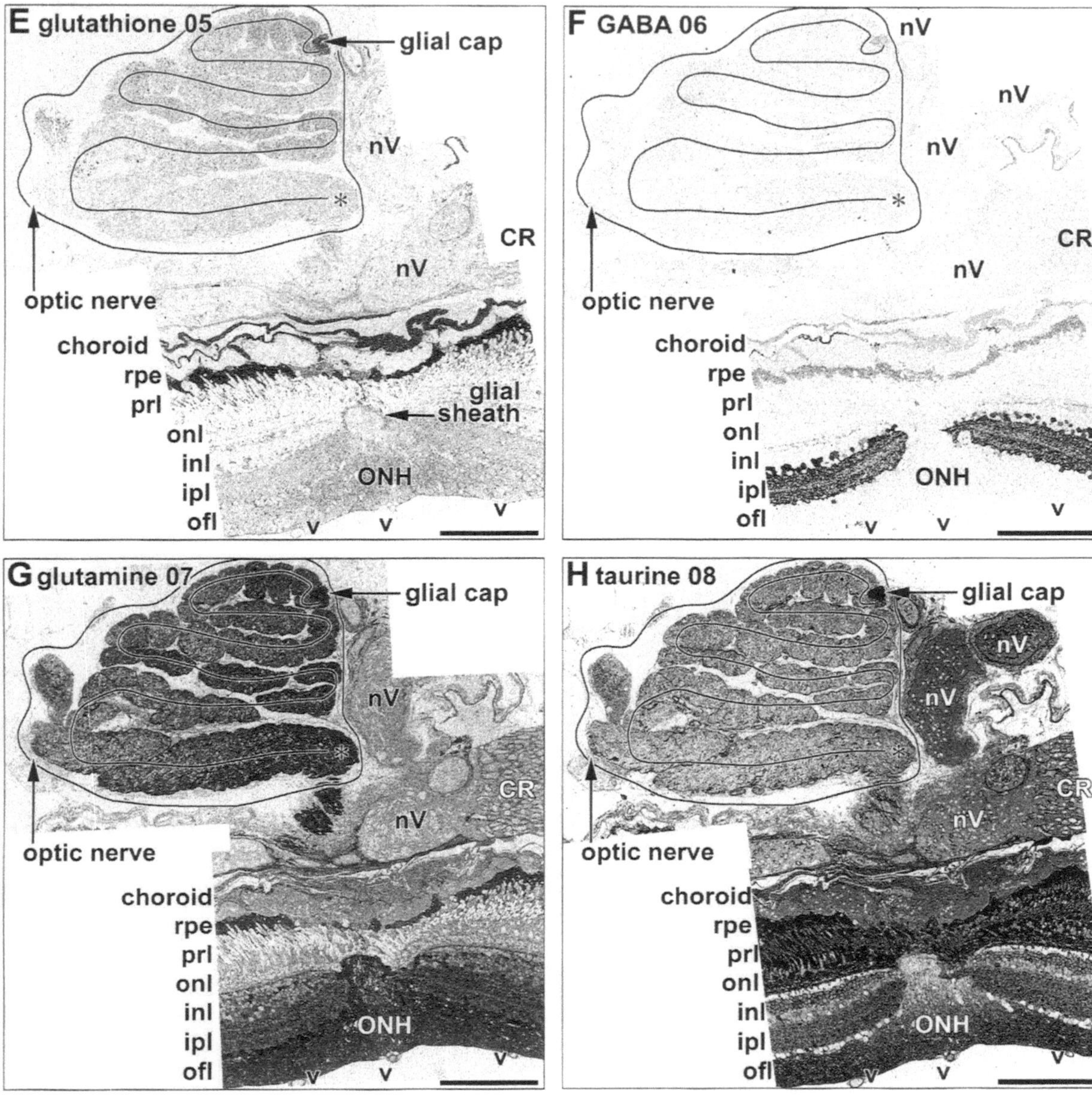

Fig. 33. A high-resolution library of the optic nerve head and folded optic nerve ribbon. A, alanine 01. B, aspartate 02. C, glutamate 03. D, glycine 04. E, glutathione 05. F, GABA 06. G, glutamine 07. H, taurine 08. Asterisk denotes the portion of the optic nerve ribbon (outlined) that arises from central retina and maps to increasingly peripheral loci (folded line) to end at the newest, most peripheral ganglion cell axons near the glial cap. Abbreviations: CR, choroidal rete; ofl, optic fiber later; ONH, optic nerve head; prl, photoreceptor layer; nV, vasculature of the optic nerve, Section thickness: 250 nm. Scale 100 μm.

horizontal cells of all non-mammalians (see review by Marc (1992)). Previous description of GABA signals in the zebrafish outer plexiform layer detected weak horizontal cell signals but did not resolve subtypes (Connaughton *et al.*, 1999). H1 horizontal cells are the only teleost horizontal cell type expressing a GABA transporter (Marc *et al.*, 1978; Marc, 1992; Verweij *et al.*, 1998; Watt *et al.*, 2000) or GAD (Lam *et al.*, 1979; Brandon, 1985; Marc, 1992). So we might expect zebrafish H1 horizontal cells to behave similarly. Color opponent horizontal cells in zebrafish and other teleost fishes lack significant GABA signals, though they do have the characteristic $\tau-$ signature of horizontal cells in general and maintain the same glutamate contents as light-adapted cones. Rod horizontal cells also lack any GABA signal, but have a characteristic $D+E+$ signal with millimolar glutamate content. The population sizes of H1, Hx and rod horizontal cells in the zebrafish retina suggest that the Hx population must be similar to that of the goldfish, which is known to harbor both H2 and H3 cone horizontal cells (*e.g.* Miyachi *et al.* (1999)). If the circuitry of zebrafishes matches that of

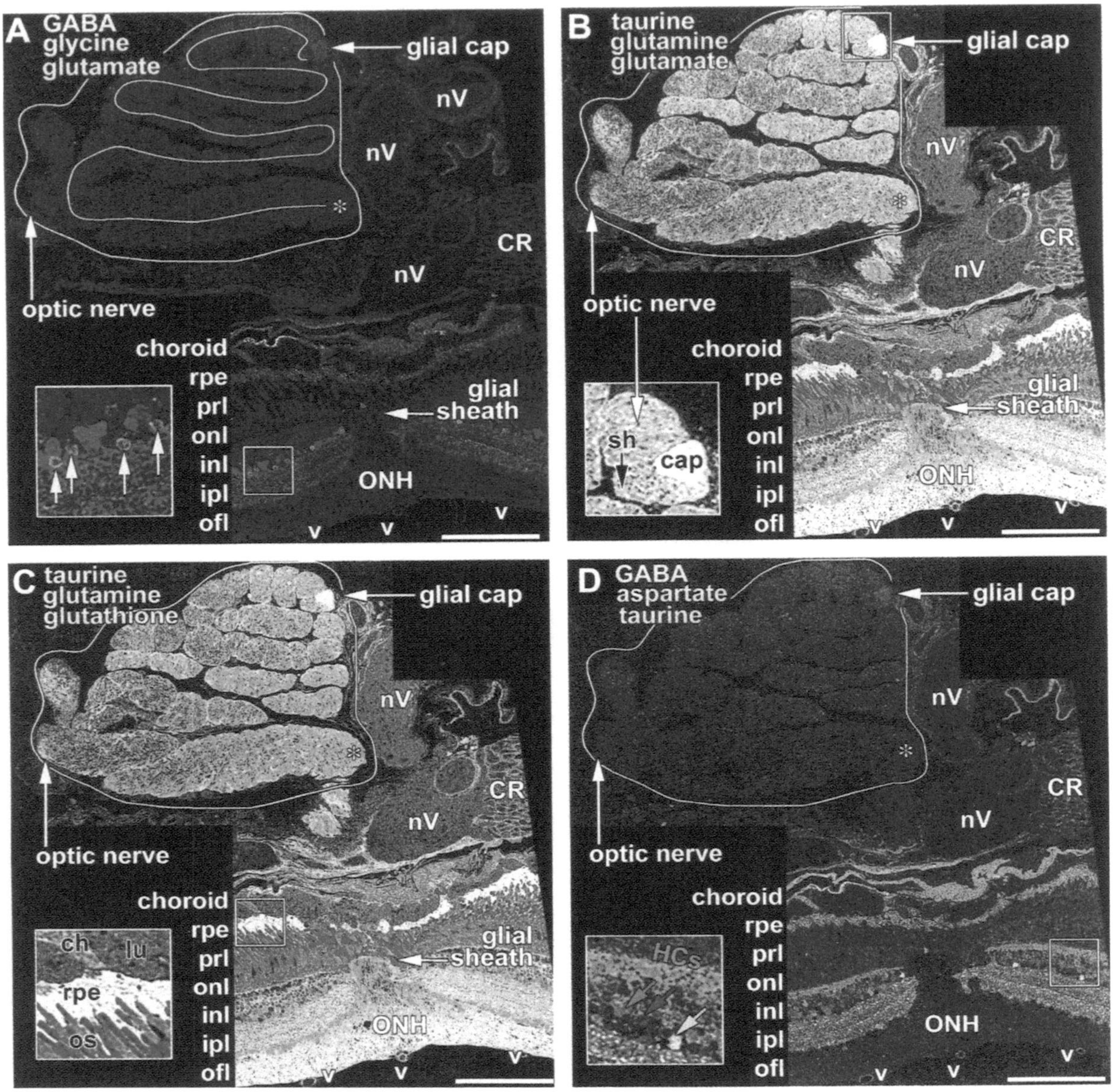

Fig. 34. Triplet mappings of the optic nerve head and folded optic nerve ribbon: A, γGE $\rightarrow$ *rgb*. The two inset boxes are 50 μm square and show the complete separation of $\gamma+$ and G+ amacrine cells. B, τQE $\rightarrow$ *rgb*. The two inset boxes are 50 μm square and show the separate signal patterns of the peripheral optic nerve glial cap (white, high τQE signal) and the normal nerve sheath glia (yellow, high τQ signal). C, τQJ $\rightarrow$ *rgb*. The two inset boxes are 50 μm square, showing the strong spectral, and hence metabolic, differentiation of rod outer segments (red), the RPE (white), choroidal endothelium (lavender), and the choriocapillaris vascular lumen (brown). D, γDτ $\rightarrow$ *rgb*. The two inset boxes are 50 μm square, showing the high D+ signal of the horizontal cell layer (green) and three aspartate-differentiated classes of amacrine cells (arrows). Scale 100 μm.

other cyprinids, both H2 and H3 horizontal cells should exist to service the green and blue cones respectively. The signatures of non-H1 horizontal cells are challenging, for they do not fit any conventional view of lateral inhibitory transmission. Marc *et al.* (1990, 1995) noted the elevated glutamate contents of rod horizontal cells and suggested that they may use glutamate rather than GABA to signal to their targets. Kamermans and his colleagues have argued that horizontal cell feedback to cones may not be exclusively chemical but rather an ephaptic signaling, perhaps involving current leaks through connexin-based hemijunctions in horizontal cell dendrites (Kamermans & Spekreijse, 1999; Kamermans *et al.*, 2001). That may solve the communication problem with cones at least, but it does not help us explain why cyprinid horizontal cells should have such distinct molecular signatures. And we do know that many horizontal cells form conventional chemical synapses with non-photoreceptor targets (Marc & Liu, 1984; Marshak & Dowling, 1987; Van Haesendonck

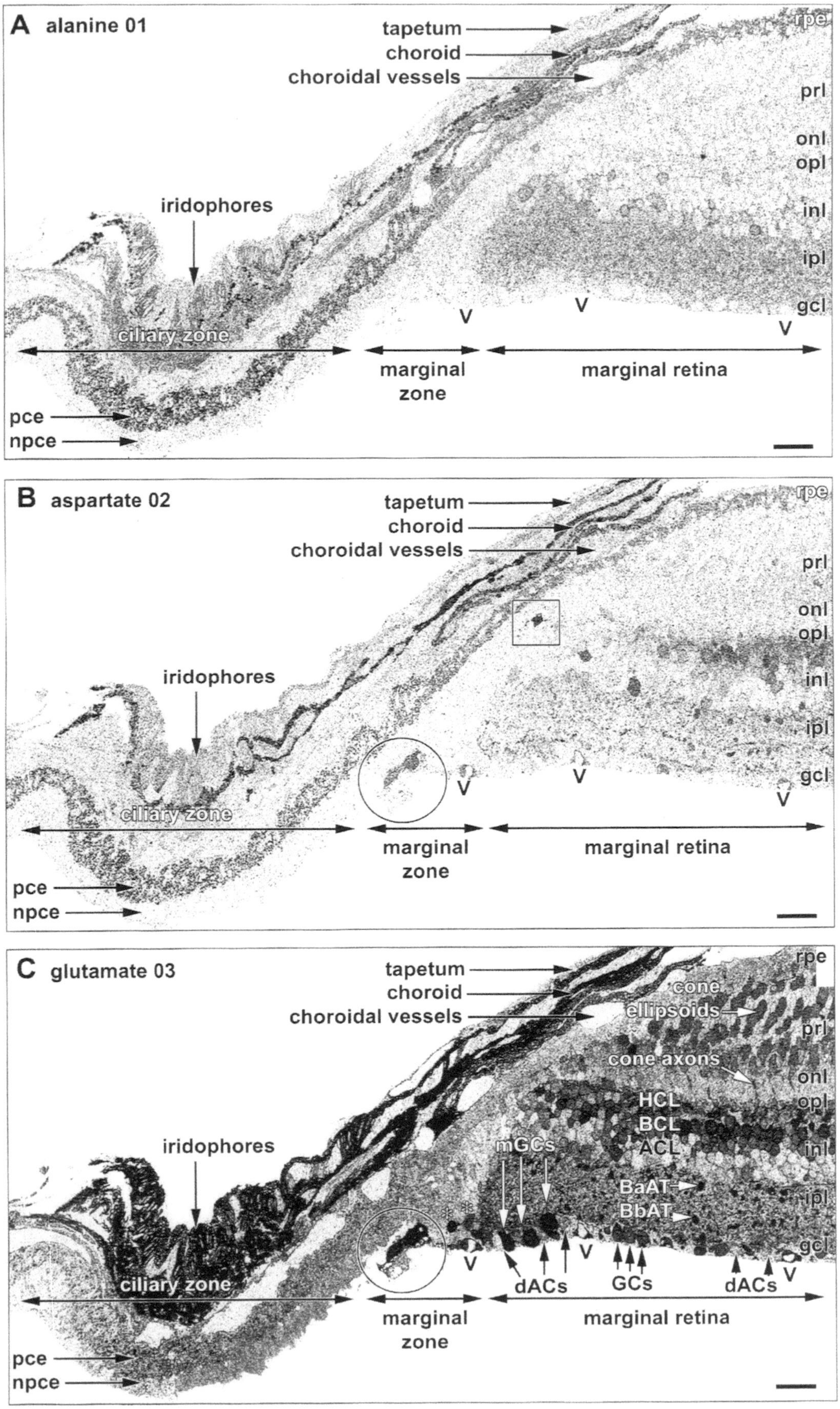

A alanine 01
tapetum
choroid
choroidal vessels
rpe
prl
onl
opl
inl
ipl
gcl
iridophores
ciliary zone
marginal zone
marginal retina
pce
npce
B aspartate 02
tapetum
choroid
choroidal vessels
rpe
prl
onl
opl
inl
ipl
gcl
iridophores
ciliary zone
marginal zone
marginal retina
pce
npce
C glutamate 03
tapetum
choroid
choroidal vessels
rpe
cone ellipsoids
prl
cone axons
onl
opl
HCL
BCL
ACL
inl
mGCs
BaAT
BbAT
ipl
gcl
iridophores
ciliary zone
dACs
GCs
dACs
marginal zone
marginal retina
pce
npce

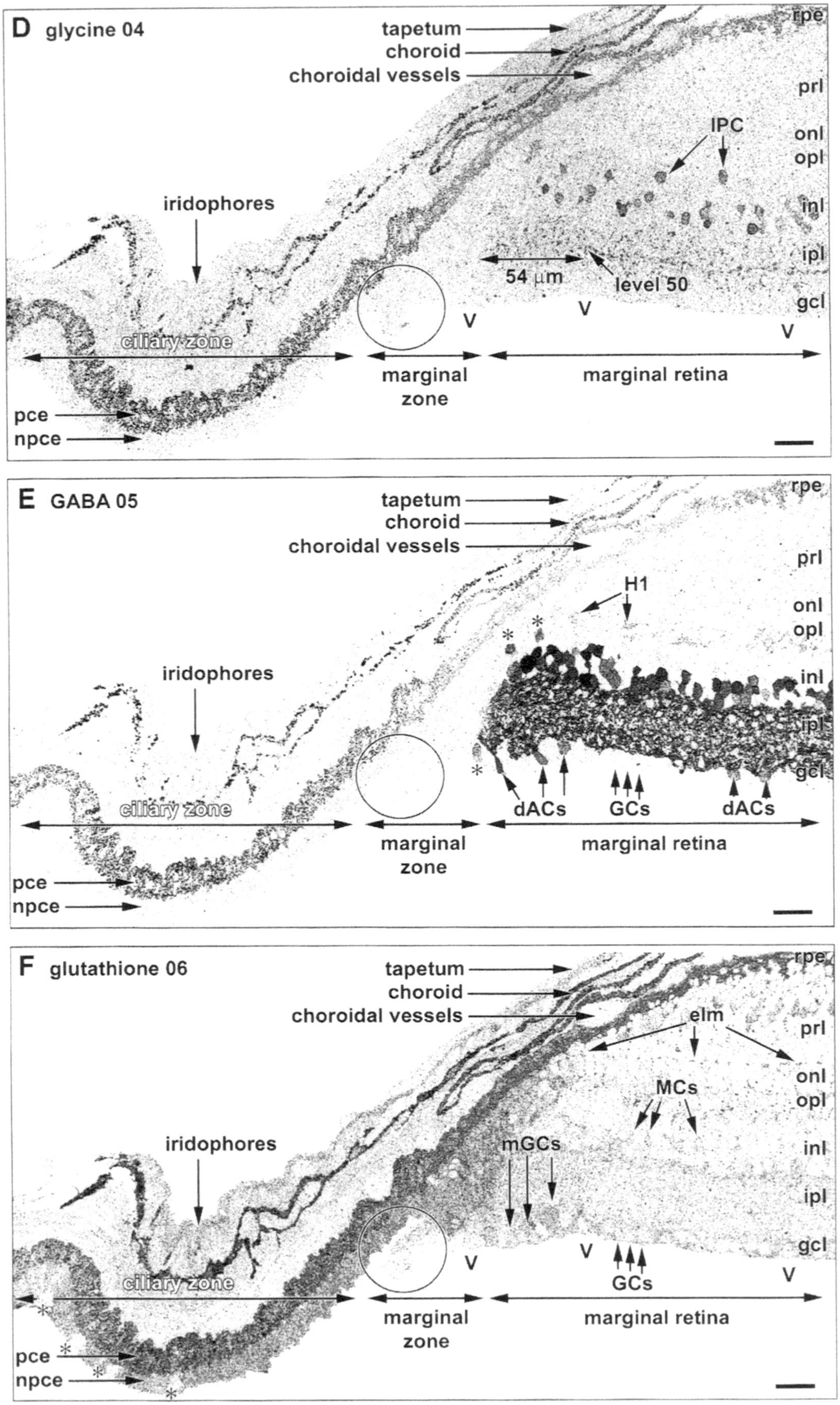
D glycine 04
tapetum
choroid
choroidal vessels
rpe
prl
IPC
onl
opl
iridophores
inl
ipl
54 μm
level 50
gcl
V
V
V
ciliary zone
marginal zone
marginal retina
pce
npce
E GABA 05
tapetum
choroid
choroidal vessels
rpe
prl
H1
onl
opl
iridophores
inl
ipl
gcl
dACs
GCs
dACs
ciliary zone
marginal zone
marginal retina
pce
npce
F glutathione 06
tapetum
choroid
choroidal vessels
rpe
elm
prl
onl
MCs
opl
iridophores
mGCs
inl
ipl
gcl
V
V
GCs
V
ciliary zone
marginal zone
marginal retina
pce
npce

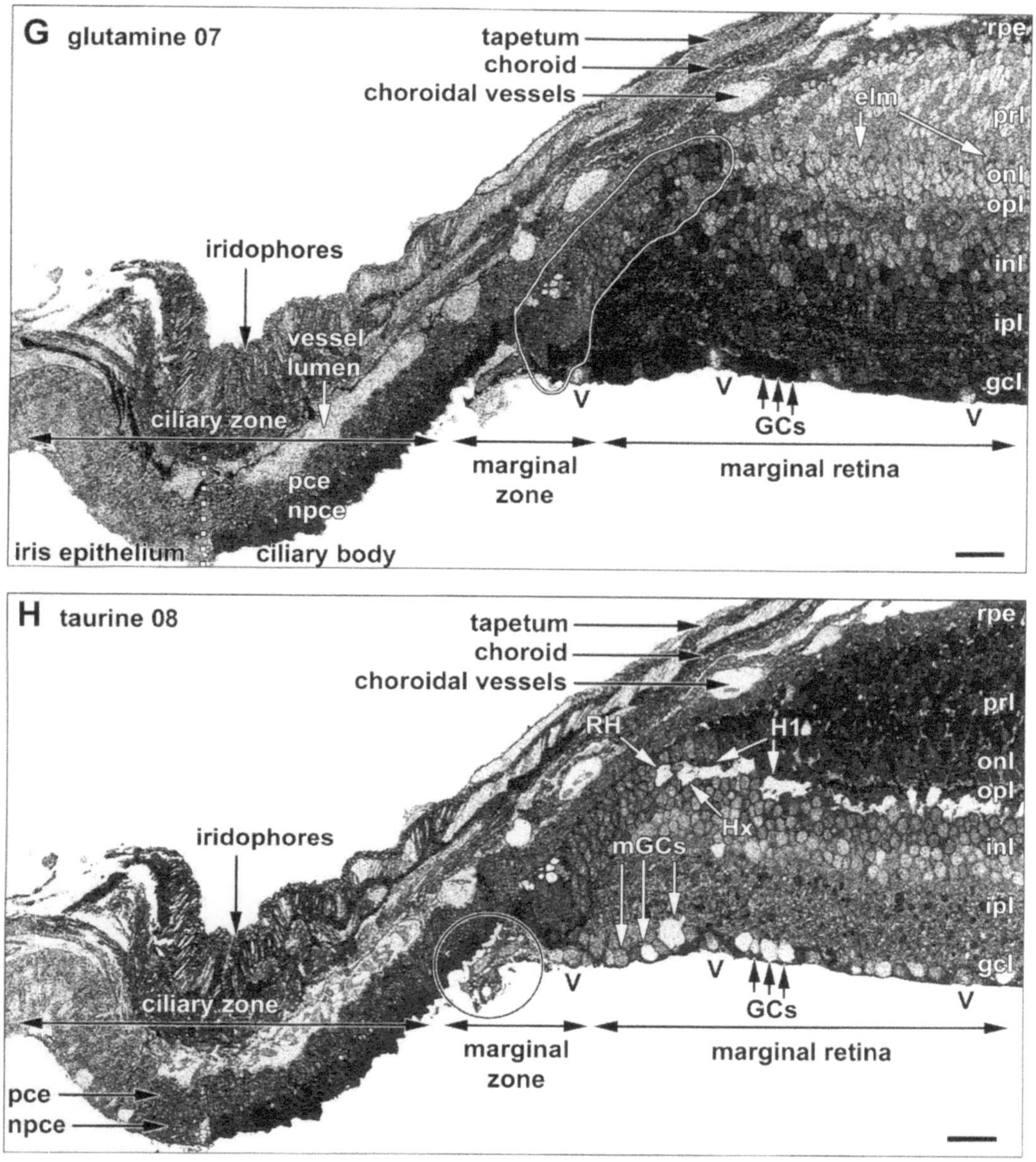

Fig. 35. A high-resolution library of the marginal retina, the marginal zone, and the ciliary zone. A, alanine 01. B, aspartate 02. Two unknown cell types display high D signals (circle), while a presumed maturing RH cell in the marginal zone is boxed. C, glutamate 03. Asterisks denote two presumed immature ganglion cells in the marginal zone. D, glycine 04. Two glycinergic interplexiform cells are marked (IPC). The characteristic level 50 glycinergic band approaches within 54 μm of the marginal zone. E, GABA 05. Asterisks denote presumed maturing conventional and displaced amacrine cells in the marginal zone. F, glutathione 06. Asterisks mark glutathione-deficient cells in the non-pigmented ciliary epithelium. G, glutamine 07. The distinctive, low glutamine marginal zone is outlined. There is a distinct molecular border (dotted) between the ciliary body and iris epithelium. H, taurine 08. Horizontal cells are clearly emerging from the marginal zone in their proper laminar position. Abbreviations: mGCs, maturing ganglion cells; npce, nonpigmented ciliary epithelium; pce, pigment ciliary epithelium; Section thickness: 250 nm. Scale 100 μm.

et al., 1993), so their signatures might still need to be coupled to their neurotransmitter function.

Another puzzling feature of teleost cone horizontal cells are their specialized axon terminal expansions, which are unbranched, nematode-shaped processes filled with microtubules (and occasional conventional synapses) that do not contact photoreceptors and course through the inner nuclear layer. The zebrafish does, in fact, have these processes but they are some 100-fold smaller in volume than those of the carp relatives and form a distinct layer just proximal to the horizontal cell layer proper. Whether they actually perform any network operations remains to be seen, and their only likely functions would seem to be lateral homotypic coupling with other horizontal cell axons or feedforward synapses onto bipolar cell proximal

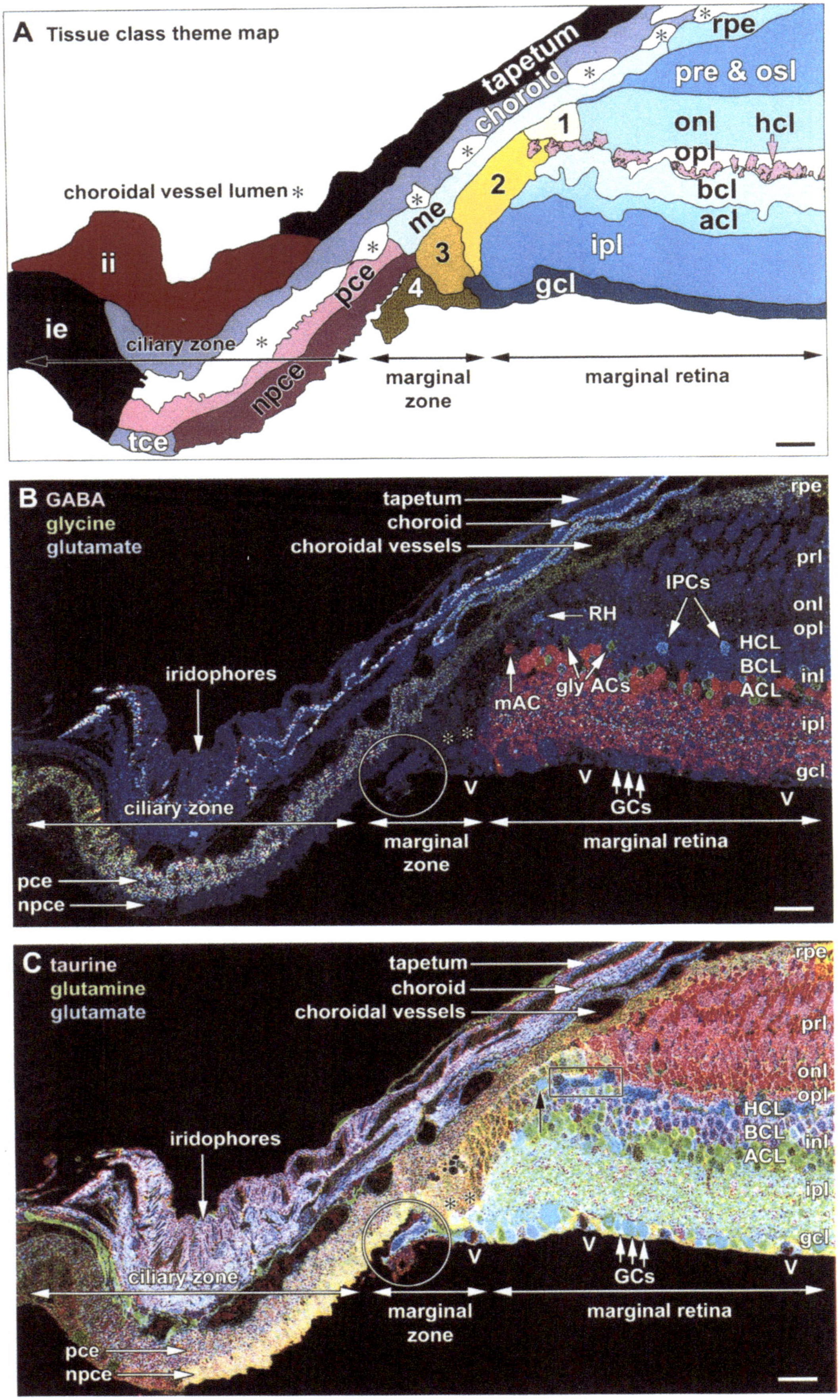
A Tissue class theme map
rpe
tapetum
choroid
pre & osl
onl
hcl
opl
bcl
acl
ipl
gcl
choroidal vessel lumen *
me
1
2
3
4
ii
ie
pce
npce
tce
ciliary zone
marginal zone
marginal retina
B GABA
glycine
glutamate
tapetum
choroid
choroidal vessels
rpe
prl
onl
opl
IPCs
RH
HCL
BCL
ACL
inl
iridophores
mAC
gly ACs
ipl
gcl
V
GCs
ciliary zone
marginal zone
marginal retina
pce
npce
C taurine
glutamine
glutamate
tapetum
choroid
choroidal vessels
rpe
prl
onl
opl
HCL
BCL
ACL
inl
iridophores
ipl
gcl
V
GCs
ciliary zone
marginal zone
marginal retina
pce
npce

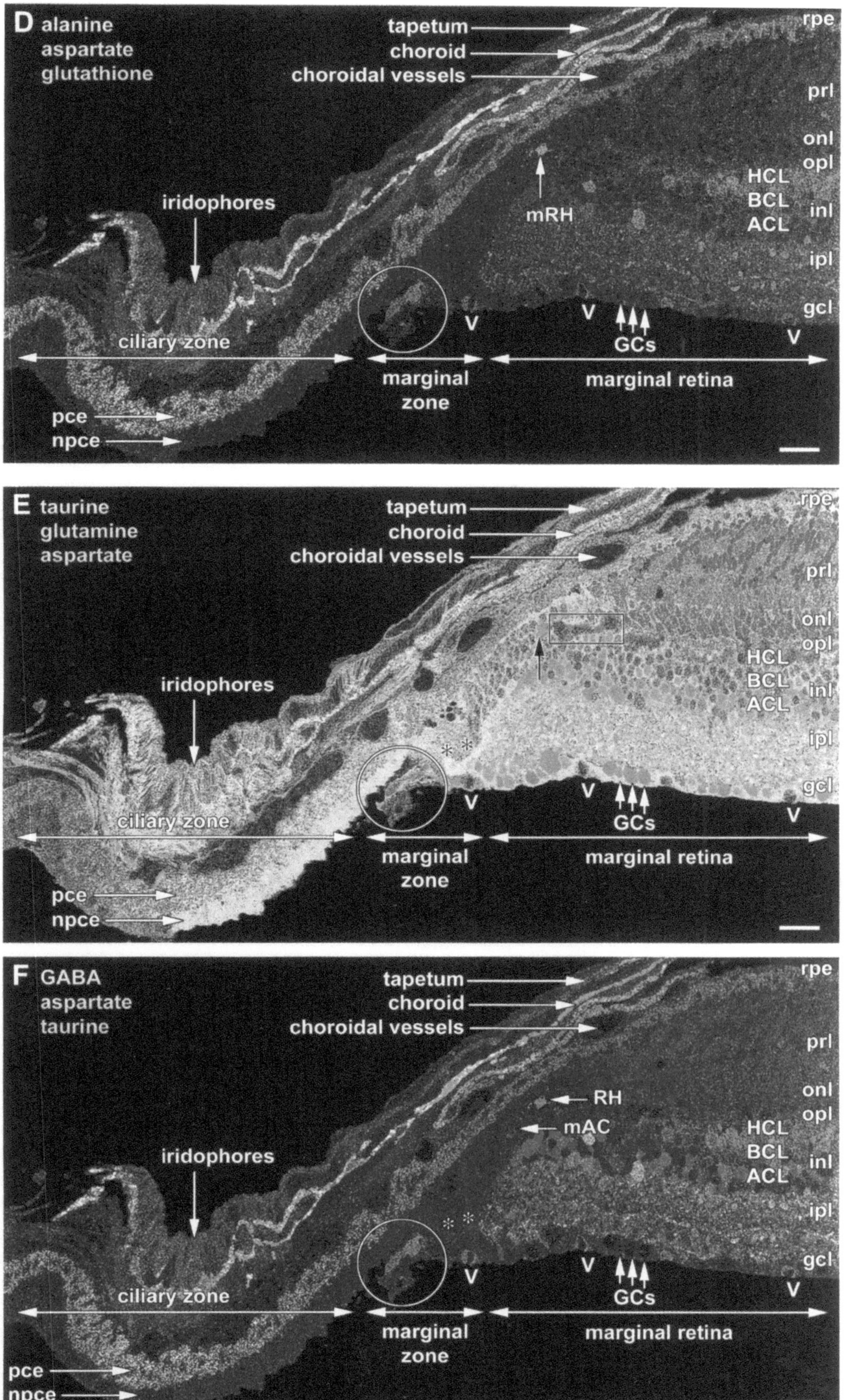
D alanine
aspartate
glutathione
tapetum
choroid
choroidal vessels
rpe
prl
onl
opl
HCL
BCL
ACL
inl
ipl
gcl
iridophores
mRH
V
GCs
ciliary zone
marginal zone
marginal retina
pce
npce
E taurine
glutamine
aspartate
tapetum
choroid
choroidal vessels
rpe
prl
onl
opl
HCL
BCL
ACL
inl
ipl
gcl
iridophores
V
GCs
ciliary zone
marginal zone
marginal retina
pce
npce
F GABA
aspartate
taurine
tapetum
choroid
choroidal vessels
rpe
prl
onl
opl
RH
mAC
HCL
BCL
ACL
inl
ipl
gcl
iridophores
V
GCs
ciliary zone
marginal zone
marginal retina
pce
npce

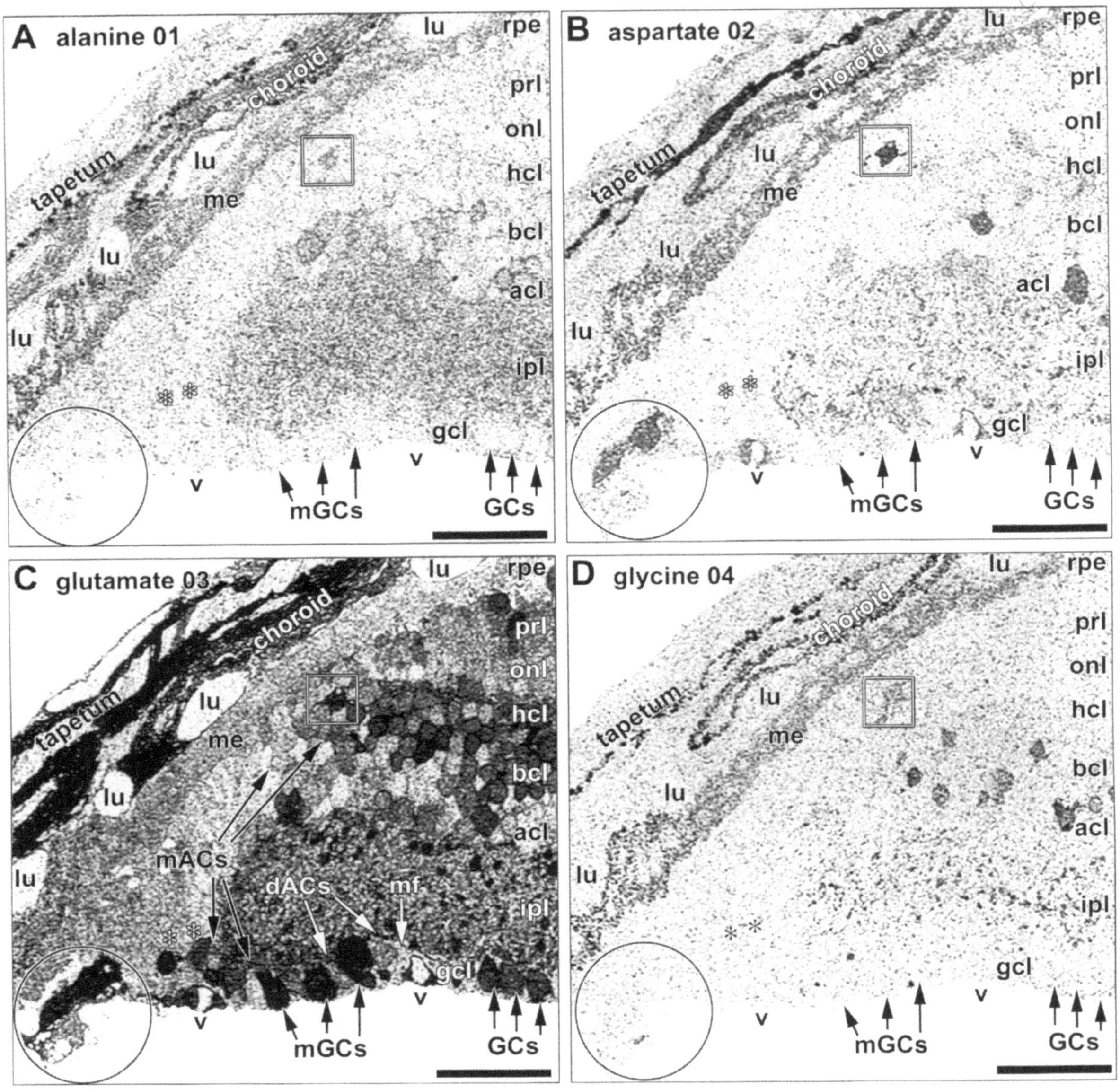

dendrites. Both H1 and Hx horizontal cells have these terminals, but zebrafish D+ horizontal cells seem to lack them. Since they are most proximal, appear to lack axon terminal expansions and have glutamate signals similar to goldfish rod horizontal cells, the D+ horizontal cells are likely to be zebrafish rod horizontal cells, and we have so treated them throughout. Teleost horizontal cells are not strictly homologous to tetrapod horizontal cells, as the latter taxa lack rod horizontal cells and instead express axons from their long-wave driven horizontal cells that contact mixed rods and cones. In non-mammalians, these are still $\gamma+$ horizontal cells. It is quite intriguing that juvenile goldfish horizontal cells are some 20-fold smaller in volume than adult horizontal cells (Raymond, 1990): they are initially like zebrafish and then become cyprinine. Mammalian horizontal cells display a complex and non-homologous pattern. Most, but not all mammals express cone horizontal cells that lack axons, but some of these to express low levels of GABA (such as central rabbit type A horizontal cells). All mammals express a cone horizontal cell with a specialized axon terminal expansion that

Fig. 36. (Previous two pages) Tissue theme map and triplet mappings of the marginal retina, the marginal zone, and the ciliary zone. A, Theme map. Four signature zones of the marginal zone are marked (1,2,3,4). B, γGE $\rightarrow$ *rgb*. C, τQE $\rightarrow$ *rgb*. The inset boxes outlines the emergent horizontal cells and the arrow denotes the RH cell. D, ADJ, the presumed emerging RH cell is indicated. E, τQD $\rightarrow$ *rgb*. F, γDτ $\rightarrow$ *rgb*. Scale 100 μm.

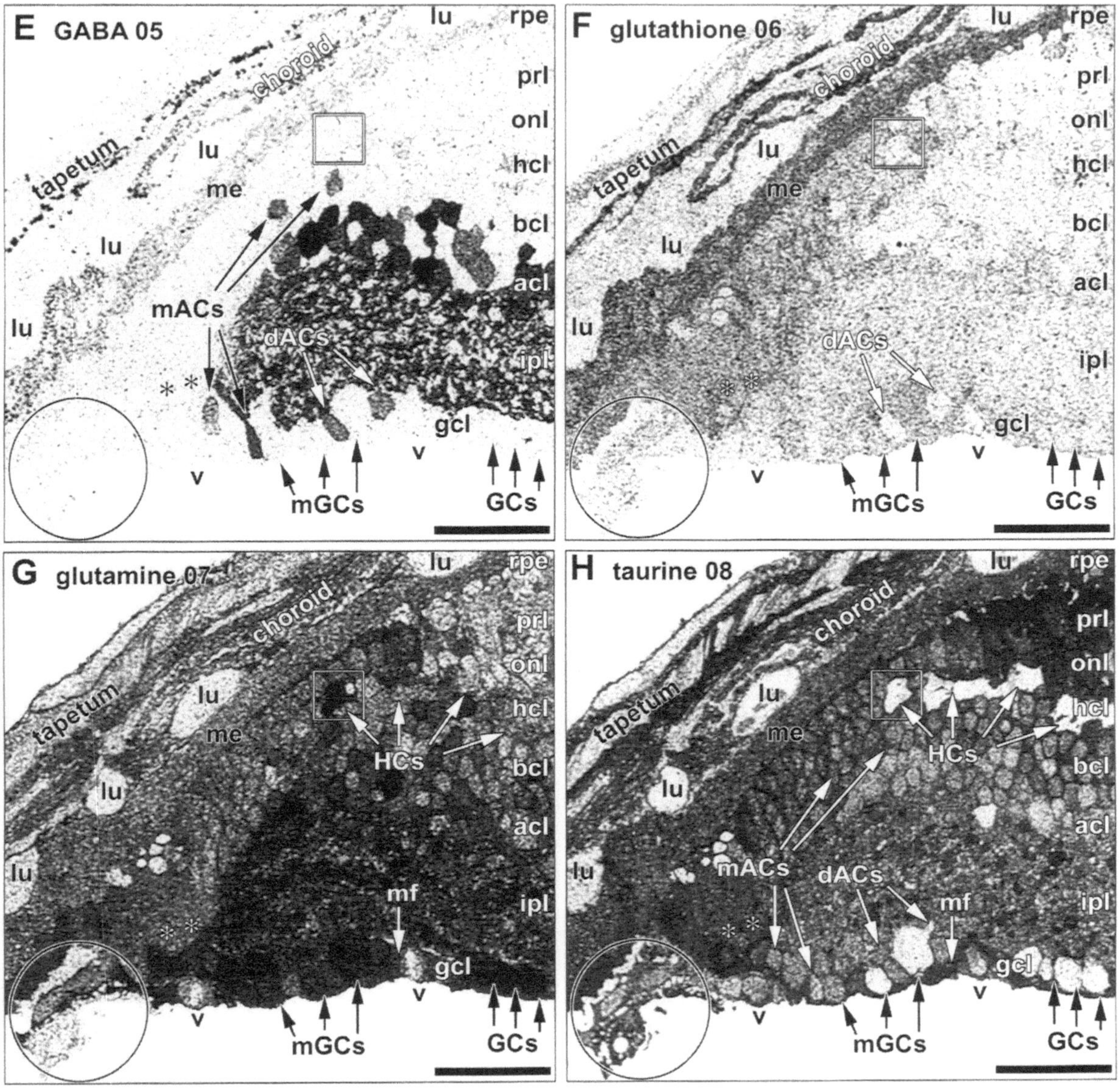

Fig. 37. A high-resolution library of the marginal zone. A, alanine 01. The maturing RH cells is boxed, the marginal zone 4 D+ cells circled and the immature ganglion cells in the marginal zone are marked with asterisks. B, aspartate 02. C, glutamate 03. D, glycine 04. E, GABA 05. F, glutathione 06. Note that the maturing ganglion cells (mGCs) have unusually high glutathione signals. G, glutamine 07. H, taurine 08. Section thickness: 250 nm. Scale 50 μm.

receives input from rods, but with few exceptions, these do not express significant GABA content. The neurochemistry of horizontal cells, after all these years, remains a puzzle.

BIPOLAR CELL CLASSES AND MOLECULAR PHENOTYPES

The bipolar cell populations of cyprinids are complex and divided into mixed rod-cone bipolar cells and pure cone bipolar cells (Stell, 1965, 1967; Scholes & Morris, 1973; Scholes, 1975; Ishida *et al.*, 1980b; Sherry & Yazulla, 1992). This complexity, however, provides a strong basis for testing homology and the zebrafish proves to be virtually identical to all other cyprinids. The rod-cone bipolar cells have been long known to be of two groups, ON-center and OFF-center, further segregated by details of photoreceptor contact patterns into 5. The zebrafish retina expresses generic cyprinid features of distinctive terminal patterning and Eτ bipolar cell signatures. Numerous subtypes were reported by Connaughton *et al.* (1999) based on qualitative examination of glutamate immunoreactivity. There are clearly at least two kinds of Ba terminals as in the goldfish retina. In goldfishes, Ba2 bipolar cells are also distinctively labeled by [^{3}H] serotonin uptake (Marc *et al.*, 1988b) and were proposed to be heterologously coupled to S1 GABAergic/serotoninergic amacrine cells. While

Table 1. Neuronal cell types, classes and signatures in the zebrafish retina.

Cell code	*Cell type*	*Molecular phenotypes*	*Natural classes*	*Primary signature*	*Primary transmitter class*
Retina		**≈25**	**≈114**		
1.x.x.x	**Vertical channels**	**≈11**	**≈38**	**E**	**E**
1.1.*x*.*x*	*Photoreceptors*	4	5	τE	E
1.1.1.x	Rod photoreceptors	1	1	τE	E
1.1.1.1	Rods	Rod	1	τE	E
1.1.2.x	Cone photoreceptors	3	4	τE	E
1.1.2.1	LD cones	LD/SD	1	τE	E
1.1.2.2	SD cones	LD/SD	1	τE	E
1.1.2.3	SS cones	SS	1	τE	E
1.1.2.4	MSS Cones	MSS	1	τE	E
1.2.x.x	*Bipolar cells*	6	18	τE	E
1.2.x.x	γ+BC	BC1	1	τEγ	γ ?
1.2.1.x	Mixed rod-cone BCs	3	5	τED	E
1.2.1.1	Bb1 BC	BC5	1	τED	E
1.2.1.2	Bb2 BC	BC5	1	τED	E
1.2.1.3	Bb3 BC	BC6	1	τED	E
1.2.1.4	Ba1 BC	BC2	1	τED	E
1.2.1.5	Ba2 BC	BC2	1	τED	E
1.2.3.x	Cone BCs	2	≈12	τE, τEQ	E
1.2.3.x	C BC	BC3,4	≈12	τE, τEQ	E
1.3.*x*.*x*	*Ganglion cells*	GC	≈15	EQ	E
2.x.x.x	**Lateral channels**	**≈14**	**≈76**	**mixed**	**γ, G, DA**
2.1.*x*.*x*	*Horizontal cells*	3	4	*mixed*	γ ?
2.1.1.x	Cone horizontal cells	2	3	mixed	γ ?
2.1.1.1	H1 cone HCs	H1	1	γ	γ ?
2.1.1.2	H2 cone HCs	Hx	1	τ−	?
2.1.1.3	H3 cone HCs	Hx	1	τ−	?
2.1.2.1	Rod HCs	RH	1	DE	?
2.2.*x*.*x*	*Amacrine cells*	11	≈72	*mixed*	γ, G, DA
2.2.1.x	γ+Amacrine cells	7	?	γ	γ
2.2.1.1	γAC1	γAC1	30%	γτQ	γ
2.2.1.2	γAC2	γAC2	6%	γτEQ	γ
2.2.1.3	γAC3	γAC3	30%	γEQ	γ
2.2.1.4	γAC4	γAC4	14%	γQ	γ
2.2.1.5	γAC5	γAC5	6%	Low γ	γ
2.2.1.6	γAC6	γAC6	10%	γEDQ	γ
2.2.1.7	γAC7	γAC7	4%	High γ	γ
2.2.2.x	G+ Amacrine cells	4	?	G	G
2.2.2.1	GAC1	GAC1 (3)	95%	G	G
2.2.2.2	GAC2	GAC2	5%	QG	G
2.3.*x*.*x*	*Interplexiform cells*		2		G
2.3.1.x	γ+Interplexiform cells		?	?	γ
2.3.2.x	G+ Interplexiform cells		1		G
2.3.1.1	G+ IPCs	GAC1	1	G	G
2.3.3.x	DA+ interplexiform cells		1		DA
2.3.3.1	DA IPCs	DA IPC	1	γ−G- DA	DA

serotonin mapping has not yet been achieved, it would be likely that the zebrafish would express a similar pattern. We also provide evidence that the bipolar cell signature class BC6 corresponds to the Bb3 bipolar cell, a cell that contacts rods and both red and green cones in the goldfish and has a wide dendritic arbor (Ishida *et al.*, 1980a). The remaining BC5 superclass is large enough to encompass at least two more classes and it is likely that it contains both Bb1 and Bb2 bipolar cells. The cone bipolar cell population is complex and likely contains some 12 classes in goldfishes. We have not been able to sort specific classes, but the pattern of cone bipolar cell terminal labeling in the inner plexiform layer is, in fact, indistinguishable from that in the goldfish. Given the same spectral classes of cones and horizontal cells in carps and zebrafishes, we find no evidence

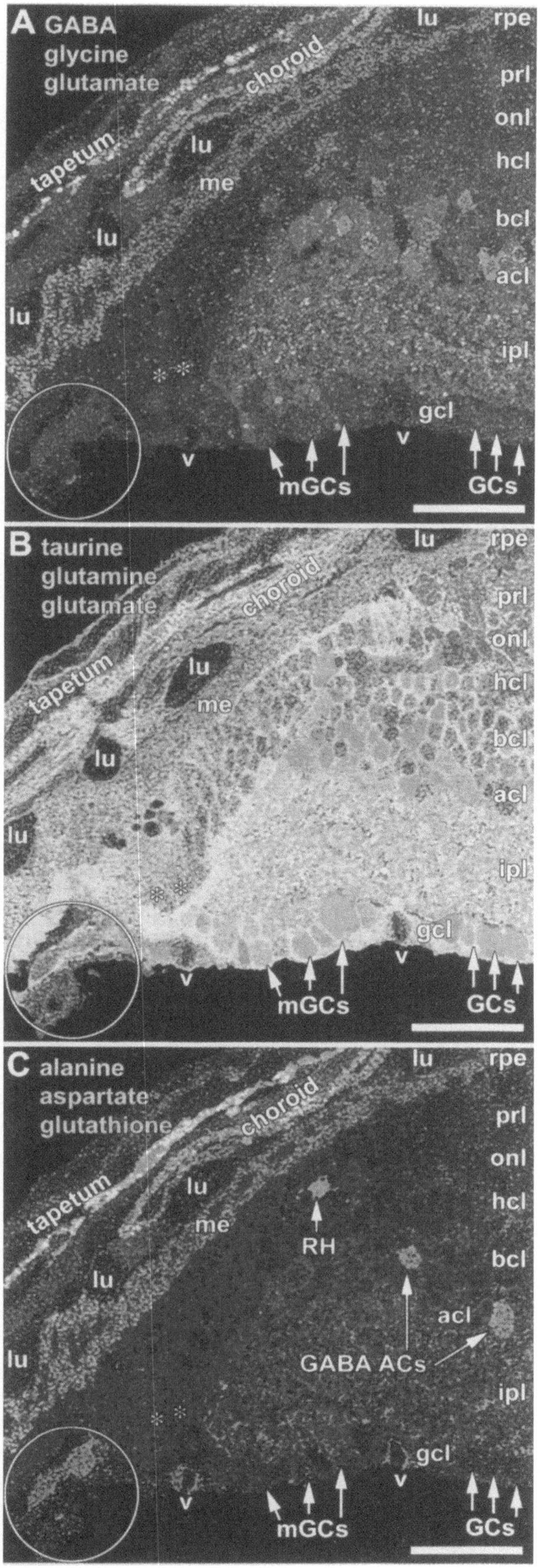

that the bipolar cell population of the zebrafish is in any way less complex. The molecular phenotype classes of bipolar cells are better differentiated in zebrafish than in carps, a phenomenon for which we have no good explanation. Class BC1 bipolar cell signatures in zebrafish are a strong departure from all other cyprinids examined so far in their weak E and modest γ signals. But for their high τ signal, one could mistake their phenotype for H1 horizontal cells. Other non-mammalians display γ+ bipolar cells (Yang & Yazulla, 1988); this is an *apparent* departure from other cyprinids. There is no rule that precludes a bipolar cell from using an inhibitory neurotransmitter, but we do know that such phenomena in the mammalian retina can be due to heterocellular coupling. The ON-center cone bipolar cells of mammals are coupled to glycinergic amacrine cells, and a signal leak is inevitable. The problem with this analogy is simply that mammalian ON-center bipolar cells have a perfectly good Eτ glutamatergic signature similar to all vertebrate bipolar cell types, including the zebrafish, but class BC1 cells in the zebrafish *do not*. Their signal more closely resembles a weak γ+ GABAergic cell. It is also likely that they represent a single natural class, as the spacing of their somas has a conformity ratio >3.

AMACRINE CELL COHORTS AND MOLECULAR PHENOTYPES

The amacrine cells of the zebrafish retina generally resemble those of the carps: separate γ+ or G+ populations, with the γ+ cells dominating the inner plexiform layer. In fact, over 90% of the neuronal space in the zebrafish inner plexiform layer is γ+, just as in goldfish. Previous reports of GABA banding in fish retinas are actually anoxia-induced and technical artefacts. The patterning of γ+ and G+ signals in the zebrafish inner plexiform layer would be indistinguishable from other cyprinids but for the distinctive band at L 50, first described by Connaughton *et al.* (1999). Since few other teleosts have yet been surveyed comprehensively, it is not difficult to tell whether this is apomorphic for the zebrafish. That aside, the sparseness of G+ terminals in the inner plexiform layer, the presence of glycinergic interplexiform cell, and the uniformity of GABA signals is strongly homologous for cyprinids in general. Breaking down the amacrine cell populations will be daunting and there is, in fact, only one quantitative assessment of amacrine cell diversity: the Golgi study of Wagner and Wagner (1988). In that work the

Fig. 38. Triplet mappings of marginal zone. A, γGE → *rgb*. B, τQE → *rgb*. C, ADJ, the presumed emerging RH cell is indicated. Scale 50 μm.

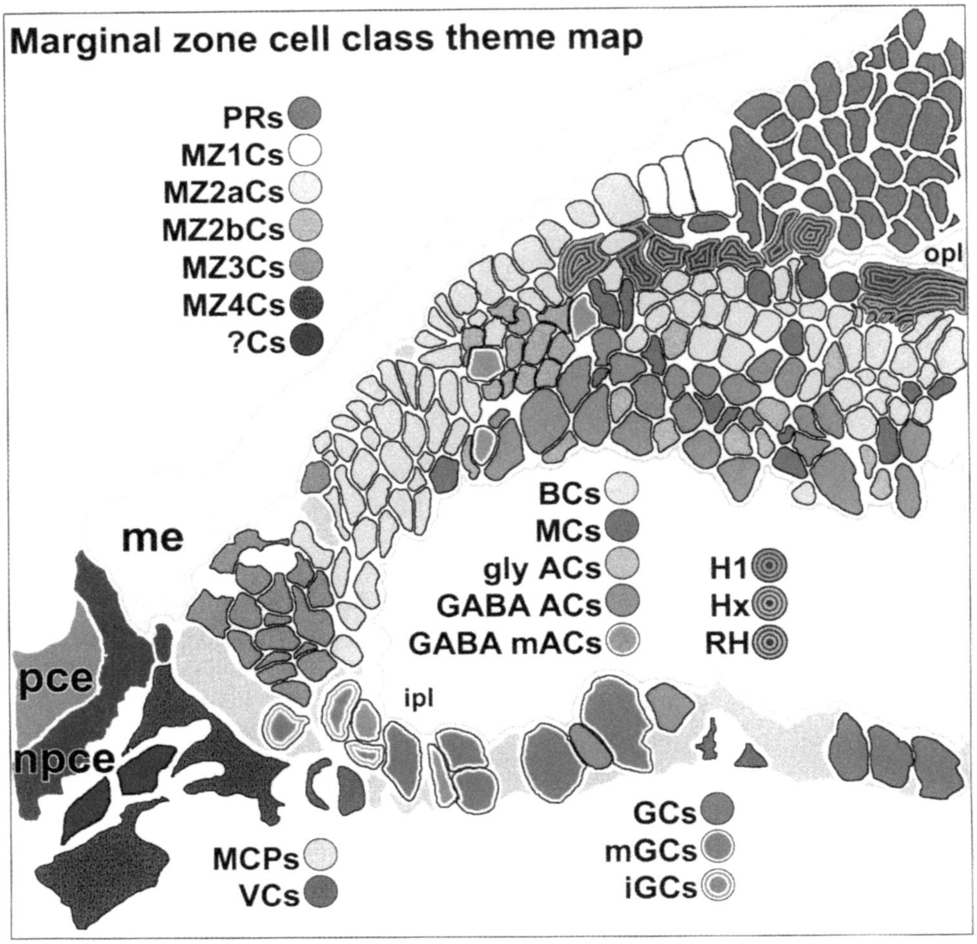

Fig. 39. A formal theme map of the marginal zone. Each cell type is coded by class membership based on isodata clustering and histogram deconvolution. Abbreviations: iGCs, immature ganglion cells; MCPs, Müller cell processes; mGCs, maturing ganglion cells; GABA mACs, maturing GABA amacrine cells; MZ1Cs, marginal zone 1 cells; MZ2aCs, marginal zone 2a cells; MZ2bCs, marginal zone 2b cells; MZ3Cs, marginal zone 3 cells; MZ4Cs, marginal zone 4 cells, PRs, photoreceptors; ?Cs, unknown cells associated with MZ4; VCs, vascular cells.

authors concluded that there were $\approx$70 distinct classes of amacrine cells. Given that the zebrafish retina contains some 30,000 amacrine cells/mm^2, that would limit the density of amacrine cell classes to an average of $\approx$400/mm^2 if all were equally frequent. Of course that is not likely to be the case. Goldfish displaced amacrine cells are γ+ cholinergic starburst homologues and have a density of around 1600/mm^2 while glycinergic and dopaminergic interplexiform cells densities in the range 40–100/mm^2. From the atlas data provided here, zebrafish displaced amacrine cells (structurally and metabolically identical to those in goldfish) have a density of $\approx$2500/mm^2 and glycinergic interplexiform cells are between 75–100/mm^2. Thus the zebrafish amacrine cells could easily comprise 70 classes, although it is clear that the zebrafish retina is more compact, with higher cell densities of all types: a logical compensation for a small eye requiring high sampling densities to track its small prey.

The strong metabolic differentiation of the zebrafish retina and improved methods permit finer discrimination of molecular phenotypes than obtained previously for goldfish and koi. Thus we would be hesitant to say that the zebrafish is qualitatively different in this regard, since signature discrimination is explicitly quantitative. Nevertheless, the variations in E, D, Q and τ signals in γ+ amacrine cells permit a strong segregation into groups that are almost certainly superclasses.

It is difficult to make direct comparisons with mammalian retinas. At least the starburst amacrine cell γ+ cholinergic cohorts are similar, but the cholinergic system in goldfish is far more complex with five cell classes: two amacrine cell classes forming each cholinergic band (radiate and starburst amacrine cells) and a bistratified cholinergic cell (Marc *et al.*, 1993). Mammalian retinas lack true serotoninergic amacrine cells and the serotonin-accumulating γ+ rod amacrine cell of the mammalian retina (Sandell & Masland, 1989;

Table 2. Non-neuronal cell types and signatures.

Tissue and cell type	*Signature*
Tapetum	
Iridophores	E
Choroid	
Pigmented epithelium	DJE
Vascular endothelium	τE
Capillary endothelium	τE
Choroidal rete	τE
RPE	
RPE	QE
Retina	
Müller cells	QJ
Astrocytes	τE
Vascular endothelium	τDE
Ciliary Zone	
PCE	QτJ
NPCE	Qτ
Transitional epithelium	EJ
Iris epithelium	Low Qτ
Marginal Zone	EJQτ mixtures
MZ1Cs	Mixtures
MZ2aCs	Mixtures
MZ2bCs	Mixtures
MZ3Cs	Mixtures
MZ4Cs	Mixtures
MZ epithelium	JτQ

Sandell *et al.*, 1989; Massey *et al.*, 1992) has no tetrapod homologue. Mammalian glycinergic rod amacrine cells (AII) are also absent from non-mammalian retinas as none of them have a pure rod bipolar cell pathway. Considering that the best amacrine cell estimates for the rabbit retina are ≈27 (MacNeil *et al.*, 1999), it is fair to say mammalian retinas are highly simplified; that the retinas of visual non-mammalians are highly advanced and specialized. To the extent that the zebrafish is chosen as a model for studying *vision*, one would have to accept this complexity as a cost of doing business, running the risk that some features of visual encoding, development and plasticity gleaned from the zebrafish may not be easily generalized.

GANGLION CELL COHORTS AND MOLECULAR PHENOTYPES

The signatures of the zebrafish ganglion cell layer are identical to those of other cyprinds, displaying only two intrinsic neurons: EQ ganglion cells and γ+ displaced amacrine cells. On occasion, some pyriform amacrine cells have misplaced somas (Marc *et al.*, 1993) and are rare intrusions. The only strong estimate of ganglion cell diversity in cyprinid retinas comes from Hitchcock and Easter (1986), and one can estimate ≈15 classes from their data. This roughly matches the diversity of bipolar cells in cyprinids and would suggest that each ganglion cell circuit requires an average of 4 unique amacrine cell classes. This is a greater complexity than any published model for receptive field organization. There is no reason to believe that the zebrafish retina is any simpler, as it seems to have the full complement of all other cyprinid neurons.

GLIAL CELL MOLECULAR PHENOTYPES

The Müller cells of the zebrafish retina have signatures similar to other vertebrates, primarily characterized by elevated glutamine and depressed glutamate contents. This is consistent with the role of Müller cell glutamine synthetase in converting transported glutamate to glutamine with high efficiency (Pow & Robinson, 1994; Pow & Crook, 1996). The zebrafish appears unexceptional in this regard. Mammals also express GAT3 GABA transporters (Johnson *et al.*, 1996), but most bony non-mammalians and all teleosts do not have any glial GABA transport (Marc, 1992). Most vertebrates also display extremely high taurine contents in their Müller cells, yielding the distinctive τQ signatures of goldfish and koi (Marc *et al.*, 1995), avians (Kalloniatis & Fletcher, 1993; Kalloniatis *et al.*, 1994; Kalloniatis & Napper, 1996), rat (Fletcher & Kalloniatis, 1996), cat (Marc *et al.*, 1998b), rabbit (Marc, 1999a), and macaque (Kalloniatis *et al.*, 1996). However, mouse Müller cells also show a modest taurine signal (Marc, unpublished). This may be due to their small size, but this is speculation.

Although not fully analyzed here, zebrafish obviously have further glial cell types in their optic fiber layers, optic nerve head and its surrounding sheath. The high τ signal of the optic fiber layer is likely an astrocyte signal, but the Müller cells clearly traverse this zone and form a lower τ band at the vitreal margin. Thus it appears that Müller cells are the sole determinants of the inner limiting membrane in zebrafish, and that there is no astrocyte involvement, as there often appears to be in mammals.

EPITHELIAL/ENDOTHELIAL COHORTS AND MOLECULAR PHENOTYPES

The choroid, tapetum, RPE, ciliary zone and various vascular elements comprise a large mixture of non-neural cells essential to ocular function. One feature of virtually all of these cells is their relatively high glutamate signals. Some of them, such as the NPCE epithelium, choroidal pigmented epithelium, and especially the radial vitreal vessel endothelium, contain glutamate signals that rival those of neurons. Many somatic transport epithelia and immune cells have millimolar glutamate contents. It appears that the basic glutamate content of such cells is archetypal; that the high glutamate contents of neurons are derivative and not novel. The high glutamate and taurine contents of the zebrafish vascular endothelia are common to the retinal vessels of all vertebrates so far examined. The fenestrated endothelia of the choroid in all vertebrates also have much lower standing amino acids contents, so there is a molecular phenotype variation that tracks endothelial type. The signatures of the zebrafish PCE and NPCE almost exactly mirror those of the mammalian retina, with significant basal levels of glutamate, glutamine, and taurine (Marc, unpublished signatures of the ciliary body). And the RPE is similar as well, except for the extremely high glutathione and glutamine levels of the zebrafish RPE. The former may be related to the high oxygen tensions created in the outer retina by the teleost choroidal rete, serving to quench the likely excessive formation of reactive oxygen species.

THE OPTIC NERVE

The optic nerves of teleosts are well-known for their unique architecture: a loose, fasciculated bundle that in many cases approximates a folded ribbon with central axons at the thicker margin of the ribbon cross-section and peripheral axons at the thinner margin. Little is known of the homoeostatic events underlying ribbon maintenance, but molecular phenotyping suggests a shift in the types of glia that comprise the stroma and sheath of the nerve, as the molecular signatures of the retinal optic fiber layer fade out in the optic nerve head and a new spectrum arises in the nerve proper. Teleosts continually add ganglion cells and axons at the retinal margin throughout life, and these new axons find their paths to the central nervous system at the thin margin of the ribbon. Molecular phenotyping consistently shows a small cap of glutathione-rich elements with a charactertistic glial or epithelial signature, quite unlike those of glutamatergic axons. Yet this structure is similar in size and location to the non-myelinated fiber zone described by Easter *et al.*(1981)for the adult goldfish optic nerve. It is possible that the nonmyelinated axons express this unusual signature, but this would require them to be different from their somatic sources.

THE MARGINAL ZONE

Also known as the germinal zone, the marginal zone is recognized as a source of new retinal neurons at the retinal margin (Johns, 1977; Marcus *et al.*, 1999). The basic signatures of marginal zone cells (MZCs) are not very dramatic, with modest glutamine and taurine signals, and low levels of most other metabolites. This suggests that MZCs are not very metabolically active. They are small, far from the choroidal rete and have scant cytoplasm, with little room for mitochondria. A cross-section of $\approx$50–60 cells abuts the retina proper. Incipient retinal neurons clearly begin to develop mature signatures while in the marginal zone and in roughly their correct locations: the outer nuclear layer, the inner nuclear layer and the ganglion cell layer. A small strand of type MZ2aCs arises from the proximal part of the small core of the marginal zone, and those cells immediately adjacent to the ganglion cell layer proper are already developing elevated glutamate signals, although their taurine signals remain high, similar to other MZCs. As ganglion cells move into the retina proper, their taurine and glutathione levels drop, and by the time they are some 50–100 μM from the margin, they develop normal $\tau-$ J$-$ and high EQ signatures of mature ganglion cells. One plausible source of τ and J signals in immature and maturing ganglion cells is through heterologous coupling with their neighbors or even retinal glia while close to the marginal zone. The marginal zone is highly structured: distal cells apposed to the marginal epithelium appear to be incipient photoreceptors (MZ1Cs); those just beneath this layer (MZ2aCs) give rise to horizontal cells, the core generates bipolar and amacrine cells (MZ2bCs), while the thin proximal strand indistinguishable from MZ2aCs generates ganglion and displaced amacrine cells. New neurons do not produce arbors until they enter the retinal proper and then never erroneously invade the marginal zone. If there are any true embryonic stem cells that have not been precommitted, it is difficult to locate them. The bottom of the marginal zone shares a signature with the NPCE and is likely the source for expansion of the ciliary zone in growth.

FINAL NOTES ON THE GENERAL METABOLISM OF NEURONS

The basis set of small molecules was chosen in part because of prior successes in differentiating cells, and in part because of the central roles each target species plays in a variety of pathways. No one species is involved in just one conversion and some, such as glutamine and glutamate, are linked in multiple cycles and epicycles as well as some 5–10 straight conversions. Thus is it not surprising that the molecular phenotypes of cells are so diverse. If anything is clear from molecular signatures, it is that no single *simple* steady-state mixture uniquely defines neuronal neurotransmitter groups. There are wide variations in glutamate content across accepted glutamatergic neurons. In general, glutamatergic photoreceptor cells in all vertebrates tend to maintain roughly 5–10$\times$ less cytosolic glutamate than bipolar and ganglion cells. And the correlative signatures of glutamatergic neurons is also diversified across types and homogenous across species: photoreceptors are low EQ, high τ; bipolar cells are high Eτ, variable Q; ganglion cells are high EQ, and τ negative. This is true of fishes, turtles, pigeons, monkeys, rabbits and mice. The high EQ signature seems to persist into the CNS (Schmolesky, Marc & Leventhal, unpublished mammalian LGN and cortex data), there might be a basic association between long axons and the ability to produce EQ signals. Short, non-spiking neurons might have less need for large Q pools but a higher local osmoregulatory challenge, thus maintaining higher τ levels. Of course there are other osmoregulatory species that we cannot detect, but the pattern remains useful across many vertebrates so far.

In general, many $\gamma+$ amacrine cells maintain glutamate levels lower than those of bipolar and ganglion cells, but this is a quantitative and not a qualitative distinction. $\gamma+$ classes AC2 and AC3 contain glutamate levels equal to or higher many bipolar and ganglion cells. Thus any ultrastructural analysis of the inner plexiform layer neuropil using glutamate as a sole signal is doomed. Multivariate signatures are essential. The source of glycine in presumed glycinergic amacrine cells has long been a mystery. The best known eukaryote pathway is through glycine hydroxymethyl transferse with serine as a substrate (Michal, 1999) but neither elevated content of this protein or the precursor serine has been achieved by immunocytochemistry. One alternative model based on constitutive accumulation aided by the glycine transporter is attractive (Pow, 1998) though not without alternatives. It is now evident that mammals and many other vertebrates can express alanine glyoxylate transaminase (Holmes & Assimos, 1998; Lee *et al.*, 1999; Li *et al.*, 1999), with glycine and pyruvate as products. While this enzyme has not been mapped in the retina, the preferential association of alanine and glycine signals is provocative.

The discovery of extensive apparent GABA$\rightarrow$ glycine coupling in the light-adapted retina was disconcerting as we had not previously attended to this issue in amacrine cells in general. A weak colocalization of GABA in a strongly G+ amacrine cell had been described for goldfish and it is very likely that the G+ cells arborizing at level 50 in the zebrafish are homologues of these cells, in spite of the apparent difference in arborization density. If these cells are coupled, then the signals ought to be reciprocal. The problem is that the leak (if that is the explanation) appears to be roughly 10% efficient: 5–10 mM in a $\gamma+$ cell produces about 0.5–1 mM in a G+ cell. Since most G+ amacrine cell glycine signals only reach about 2–3 mM, the resultant coupling signal would be very close to the basal glycine signal of most cells and probably undetectable. Even so, this finding unexpectedly prompts reevaluation of the goldfish amacrine cell populations as a function of adaptation. Amacrine-amacrine coupling has been long known in teleosts (Naka & Christensen, 1981; Marc *et al.*, 1988a; Miyachi *et al.*, 1999), and gap junctions are widespread in the goldfish inner plexiform layer (Marc *et al.*, 1988a). None of these prior data suggested any heterotypic coupling, although sensitive multi-cell recordings in catfish led to a model where extensive heterotypic coupling was proposed (Sakai & Naka, 1988). Finally, the rare BC1 $\gamma+$ bipolar cell might be deemed the one apomorphic feature of the zebrafish but for the fact that it would be easily missed amongst the hundreds of $\gamma+$ horizontal cell axon terminals in the goldfish inner plexiform layer. The zebrafish inner nuclear layer is much better laminated than any other cyprinid and the horizontal cell axon terminals are miniscule. A careful τ mask in the goldfish might reveal these $\gamma+$ bipolar cells. In the end, the only feature that seems to distinguish the zebrafish retina from other cyprinids is its more precise laminar and spatial patterning. This argues that most of the detailed anatomy, neurochemistry and physiology performed on other cyprinids (and teleosts in general) might be safely assumed.

Note

1. Disclosure: REM is a principal of Signature Immunologics Inc.

References

AMMERMULLER, J. & WEILER, R. (1981) The ramification pattern of amacrine cells within the inner plexiform layer of the carp retina. *Cell & Tissue Research* **220**, 699–723.

BALL, G. & HALL, D. (1967) Clustering technique for summarizing multivariate data. *Behavioral Science* **12**, 153–155.

BRANDON, C. (1985) Retinal GABA neurons: Localization in vertebrate species using an antiserum to rabbit brain glutamate decarboxylase. *Brain Research* **344**, 286–295.

CAMERON, D. A. (2000) Cellular proliferation and neurogenesis in the injured retina of adult zebrafish. *Visual Neuroscience* **17**, 789–797.

CAMERON, D. A. & CARNEY, L. H. (2000) Cell mosaic patterns in the native and regenerated inner retina of zebrafish: Implications for retinal assembly. *Journal of Comparative Neurology* **416**, 356–367.

CAMERON, D. A. & POWERS, M. K. (2000) Morphology and visual pigment content of photoreceptors from injured goldfish retina. *Visual Neuroscience* **17**, 623–630.

CARROLL, R. (1988) *Vertebrate Paleontology and Evolution*. New York: W. H. Freeman and Company.

CONNAUGHTON, V. P., BEHAR, T. N., LIU, W. L. & MASSEY, S. C. (1999) Immunocytochemical localization of excitatory and inhibitory neurotransmitters in the zebrafish retina. *Visual Neuroscience* **16**, 483–490.

COOK, J. E. (1996) Spatial properties of retinal mosaics: An empirical evaluation of some existing measures. *Visual Neuroscience* **13**, 15–30.

DJAMGOZ, M. B. A., WAGNER, H. J. & WITKOVSKY, P. (1995) Photoreceptor-horizontal cell connectivity, synaptic transmission and neuromodulation. In *Neurobiology and Clinical Aspects of the Outer Retina* (edited by Djamgoz, M., Archer, S. & Vallerga, S.) pp. 155–193. London: Chapman and Hall.

EASTER, S. S., RUSOFF, A. C. & KISH, P. E. (1981) The growth and organization of the optic nerve and tract in juvenile and adult goldfish. *Journal of Neuroscience* **1**, 793–811.

FAMIGLIETTI, E. V. JR., KANEKO, A. & TACHIBANA, M. (1975) Neuronal architecture of on and off pathways to ganglion cells in carp retina. *Science* (Washington DC) **198**, 1267–1269.

FLETCHER, E. L. & KALLONIATIS, M. (1996) Neurochemical architecture of the normal and degenerating rat retina. *Journal of Comparative Neurology* **376**, 343–360.

FONNER, D. B., HOFFERT, J. R. & FROMM, P. O. (1973) The importance of the counter current oxygen multiplier mechanism in maintaining retinal function in the teleost. *Comparative Biochemistry & Physiology A—Comparative Physiology* **46**, 559–567.

HITCHCOCK, P. F. & EASTER, S. S. (1986) Retinal ganglion cells in goldfish: A qualitative classification into four morphological types, and a quantitative study of the development of one of them. *Journal of Neuroscience* **6**, 1037–1050.

HOLMES, R. P. & ASSIMOS, D. G. (1998) Glyoxylate synthesis, and its modulation and influence on oxalate synthesis. *Journal of Urology* **160**, 1617–1624.

INSELBERG, A. & DIMSDALE, B. (1990) Parallel coordinates: A tool for visualizing multidimensional geometry. *Proceedings of the First IEEE Conference on Visualization* **1**, 361–375.

ISHIDA, A., STELL, W. & LIGHTFOOT, D. (1980a) Rod and cone inputs to bipolar cells in goldfish retina. *Journal of Comparative Neurology* **191**, 315–335.

ISHIDA, A. T., STELL, W. K. & LIGHTFOOT, D. O. (1980b) Rod and cone inputs to bipolar cells in goldfish retina. *Journal of Comparative Neurology* **191**, 315–335.

JOHNS, P. R. (1977) Growth of the adult goldfish eye. III. Source of the new retinal cells. *Journal of Comparative Neurology* **176**, 343–357.

JOHNSON, J., CHEN, T. K., RICKMAN, D. W., EVANS, C. & BRECHA, N. C. (1996) Multiple gamma-Aminobutyric acid plasma membrane transporters (GAT-1, GAT-2, GAT-3) in the rat retina. *Journal of Comparative Neurology* **375**, 212–224.

JONES, B. W., HOWARD, J., BEG, A., WATT, C. B. & MARC, R. E. (1999) Ionotropic glutamatergic drive histories of amacrine cell layer neurons reported by 1-amino-4-guanidobutane (AGB) *in vivo*. *Investigative Ophthalmology and Visual Science* **40**, S440.

KALLONIATIS, M. & FLETCHER, E. L. (1993) Immunocytochemical localization of the amino acid neurotransmitters in the chicken retina. *Journal of Comparative Neurology* **336**, 174–193.

KALLONIATIS, M. & MARC, R. E. (1990) Interplexiform cells of the goldfish retina. *Journal of Comparative Neurology* **297**, 340–358.

KALLONIATIS, M., MARC, R. E. & MURRY, R. F. (1996) Amino acid signatures in the primate retina. [erratum appears in *J Neurosci* 1997 Jan 1;**17**(1), 500–503]. *Journal of Neuroscience* **16**, 6807–6829.

KALLONIATIS, M. & NAPPER, G. A. (1996) Glutamate metabolic pathways in displaced ganglion cells of the chicken retina. *Journal of Comparative Neurology* **367**, 518–536.

KALLONIATIS, M. & TOMISICH, G. (1999) Amino acid neurochemistry of the vertebrate retina. *Progress in Retinal & Eye Research* **18**, 811–866.

KALLONIATIS, M., TOMISICH, G. & MARC, R. E. (1994) Neurochemical signatures revealed by glutamine labeling in the chicken retina. *Visual Neuroscience* **11**, 793–804.

KAMERMANS, M., FAHRENFORT. I., SCHULTZ, K., JANSSEN-BIENHOLD, U., SJOERDSMA, T. & WEILER, R. (2001) Hemichannel-mediated inhibition in the outer retina. *Science* **292**, 1178–1180.

KAMERMANS, M. & SPEKREIJSE, H. (1999) The feedback pathway from horizontal cells to cones. A mini review with a look ahead. *Vision Research* **39**, 2449–2468.

LAM, D. M., SU, Y. Y., SWAIN, L., MARC, R. E., BRANDON, C. & WU, J. Y. (1979) Immunocytochemical localisation of L-glutamic acid decarboxylase in the goldfish retina. *Nature* **278**, 565–567.

LEE, I. S., NISHIKIMI, M., INOUE, M., MURAGAKI, Y. & OOSHIMA, A. (1999) Specific expression of alanine-glyoxylate aminotransferase 2 in the epithelial cells of Henle's loop. *Nephron* **83**, 184–185.

LI, X. M., SALIDO, E. C. & SHAPIRO, L. J. (1999) The mouse alanine: Glyoxylate aminotransferase gene (Agxt1): Cloning, expression, and mapping to chromosome 1. *Somatic Cell & Molecular Genetics* **25**, 67–77.

MACNEIL, M. A., HEUSSY, J. K., DACHEUX, R. F., RAVIOLA, E. & MASLAND, R. H. (1999) The shapes and numbers of amacrine cells: Matching of photofilled with Golgi-stained cells in the rabbit retina and comparison with other mammalian species. *Journal of Comparative Neurology* **413**, 305–326.

MARC, R., LI, H.-B., KALLONIATIS, M. & ARNOLD, J. (1993) Cholinergic subsets of GABAergic amacrine cells in the goldfish retina. *Investigative Ophthalmology and Visual Science* **34**, S1061.

MARC, R. E. (1982) Chromatic organization of the retina. In *Cellular Aspects of the Eye* (edited by McDevitt, D.) pp. 435–473. NY: Academic Press.

MARC, R. E. (1986) Neurochemical stratification in the inner plexiform layer of the vertebrate retina. *Vision Research* **26**, 223–238.

MARC, R. E. (1992) Structural organization of GABAergic circuitry in ectotherm retinas. *Progress in Brain Research* **90**, 61–92.

MARC, R. E. (1999a) Mapping glutamatergic drive in the vertebrate retina with a channel-permeant organic cation. *Journal of Comparative Neurology* **407**, 47–64.

MARC, R. E. (1999b) The structure of vertebrate retinas. In *The Retinal Basis of Vision* (edited by Toyoda, J.-I., Murkami, M., Kaneko, A. & Saito, T.) pp. 3–19. Amsterdam: Elsevier.

MARC, R. E. & JONES, B. W. (2002) Molecular phenotyping of retinal ganglion cells. *Journal of Neuroscience* **22**, 413–427.

MARC, R. E. & LAM, D. M. K. (1981) Glycinergic pathways in the goldfish retina. *Journal of Neuroscience* **1**, 152–165.

MARC, R. E. & LIU, W. L. (1984) Horizontal cell synapses onto glycine-accumulating interplexiform cells. *Nature* **312**, 266–269.

MARC, R. E. & LIU, W. (2000) Fundamental GABAergic amacrine cell circuitries in the retina: Nested feedback, concatenated inhibition, and axosomatic synapses. *Journal of Comparative Neurology* **425**, 560–582.

MARC, R. E., LIU, W. L., KALLONIATIS, M., RAIGUEL, S. F. & VAN HAESENDONCK, E. (1990) Patterns of glutamate immunoreactivity in the goldfish retina. *Journal of Neuroscience* **10**, 4006–4034.

MARC, R. E., LIU, W. L. & MULLER, J. F. (1988a) Gap junctions in the inner plexiform layer of the goldfish retina. *Vision Research* **28**, 9–24.

MARC, R. E., LIU, W. L., SCHOLZ, K. & MULLER, J. F. (1988b) Serotonergic and serotonin-accumulating neurons in the goldfish retina. *Journal of Neuroscience* **8**, 3427–3450.

MARC, R. E., MURRY, R. F. & BASINGER, S. F. (1995) Pattern recognition of amino acid signatures in retinal neurons. *Journal of Neuroscience* **15**, 5106–5129.

MARC, R. E., MURRY, R. F., FISHER, S. K., LINBERG, K. A. & LEWIS, G. P. (1998a) Amino acid signatures in the detached cat retina. *Investigative Ophthalmology & Visual Science* **39**, 1694–1702.

MARC, R. E., MURRY, R. F., FISHER, S. K., LINBERG, K. A., LEWIS, G. P. & KALLONIATIS, M. (1998b) Amino acid signatures in the normal cat retina. *Investigative Ophthalmology & Visual Science* **39**, 1685–1693.

MARC, R. E., STELL, W. K., BOK, D. & LAM, D. M. (1978) GABAergic pathways in the goldfish retina. *Journal of Comparative Neurology* **182**, 221–244.

MARC, R. E. & SPERLING, H. G. (1976) Color receptor identities of goldfish cones. *Science* **191**, 487–489.

MARC, R. E. & SPERLING, H. G. (1977) The chromatic organization of the goldfish cone mosaic. *Vision Research* **16**, 1211–1224.

MARCUS, R. C., DELANEY, C. L., & EASTER, S. S., JR. (1999) Neurogenesis in the visual system of embryonic and adult zebrafish (Danio rerio). *Visual Neuroscience* **16**, 417–424.

MARSHAK, D. W. & DOWLING, J. E. (1987) Synapses of cone horizontal cell axons in goldfish retina. *Journal of Comparative Neurology* **256**, 430–443.

MARSHAK, D. W., YAMADA, T. & STELL, W. K. (1982) Synaptic contacts of somatostatin immuno reactive amacrine cells in goldfish retina. *Journal of Comparative Neurology* **225**, 44–52.

MASSEY, S. C., MILLS, S. L. & MARC, R. E. (1992) All indoleamine-accumulating cells in the rabbit retina contain GABA. *Journal of Comparative Neurology* **322**, 275–291.

MICHAL, G. (1999) Amino acids and derivatives. In *Biochemical Pathways* (edited by Michal, G.) pp. 46–67. NY: Wiley & Sons.

MIYACHI, E., HIDAKA, S. & MURKAMI, M. (1999) Electrical couplings of retinal neurons. In (edited by Toyoda, J.-I., Murakami, M., Kaneko, A. & Saito, T.) pp. 171–184. Amsterdam: Elsevier.

MOECKEL, G. W., LAI, L. W., GUDER, W. G., KWON, H. M. & LIEN, Y. H. (1997) Kinetics and osmoregulation of Na^+- and Cl^--dependent betaine transporter in rat renal medulla. *American Journal of Physiology* **272**, F100–106.

MORIMURA, H., SHIMADA, S., OTORI, Y., SAISHIN, Y., YAMAUCHI, A., MINAMI, Y., INOUE, K., ISHIMOTO, I., TANO, Y. & TOHYAMA, M. (1997) The differential osmoregulation and localization of taurine transporter mRNA and Na^+/myo-inositol cotransporter mRNA in rat eyes. *Brain Research Molecular Brain Research* **44**, 245–252.

MOSINGER, J. L. & YAZULLA, S. (1985) Colocalization of GAD-like immunoreactivity and ^{3}H-GABA uptake in amacrine cells of rabbit retina. *Journal of Comparative Neurology* **240**, 396–406.

MOSINGER, J. L., YAZULLA, S. & STUDHOLME, K. M. (1986) GABA-like immunoreactivity in the vertebrate retina: A species comparison. *Experimental Eye Research* **42**, 631–644.

NAKA, K. I. & CHRISTENSEN, B. N. (1981) Direct electrical connections between transient amacrine cells in the catfish retina. *Science* **214**, 462–464.

POW, D. V. (1998) Transport is the primary determinant of glycine content in retinal neurons. *Journal of Neurochemistry* **70**, 2628–2636.

POW, D. V. & CROOK, D. K. (1996) Direct immunocytochemical evidence for the transfer of glutamine from glial cells to neurons: Use of specific antibodies directed against the d-stereoisomers of glutamate and glutamine. *Neuroscience* **70**, 295–302.

POW, D. V. & ROBINSON, S. R. (1994) Glutamate in some retinal neurons is derived solely from glia. *Neuroscience* **60**, 355–366.

RAYMOND, P. A. (1990) Horizontal cell axon terminals in growing goldfish. *Experimental Eye Research* **51**, 675–683.

RAYMOND, P. A., BARTHEL, L. K., ROUNSIFER, M. E., SULLIVAN, S. A. & KNIGHT, J. K. (1993) Expression of rod and cone visual pigments in goldfish and zebrafish: A rhodopsin-like gene is expressed in cones. *Neuron* **10**, 1161–1174.

ROBERTS, M. F. (2000) Osmoadaptation and osmoregulation in archaea. *Frontiers in Bioscience* **5**, D796–D812.

ROBINSON, J., SCHMITT, E. A., HAROSI, F. I., REECE, R. J. & DOWLING, J. E. (1993) Zebrafish ultraviolet visual pigment: Absorption spectrum, sequence, and localization. *Proceedings of the National Academy of Sciences of the United States of America* **90**, 6009–6012.

SAKAI, H. M. & NAKA, K. (1988) Dissection of the neuron network in the catfish inner retina. II. Interactions between ganglion cells. *Journal of Neurophysiology* **60**, 1568–1583.

SANDELL, J. H. & MASLAND, R. H. (1989) Indoleamine accumulation by retinal neurons exposed to blood. *Histochemistry* **92**, 57–60.

SANDELL, J. H., MASLAND, R. H., RAVIOLA, E. & DACHEUX, R. F. (1989) Connections of indoleamine-accumulating cells in the rabbit retina. *Journal of Comparative Neurology* **283**, 303–313.

SCHAFFER, S., TAKAHASHI, K. & AZUMA, J. (2000) Role of osmoregulation in the actions of taurine. *Amino Acids* **19**, 527–546.

SCHOLES, J. H. (1975) Colour receptors, and their synaptic connexions, in the retina of a cyprinid fish. *Philosophical Transactions of the Royal Society of London—Series B: Biological Sciences* **270**, 61–118.

SCHOLES, J. & MORRIS, J. (1973) Receptor—bipolar connectivity patterns in fish retina. *Nature* **241**, 52–54.

SHERRY, D. M. & YAZULLA, S. (1992) Goldfish bipolar cells and axon terminal patterns: A Golgi study. *Journal of Comparative Neurology.*

STELL, W. K. (1965) Correlation of retinal cytoarchitecture and ultrastructure in Golgi preparations. *Anatomical Record* **153**, 389–397.

STELL, W. K. (1967) The structure and relationships of horizontal cells and photoreceptor-bipolar synaptic complexes in goldfish retina. *American Journal of Anatomy* **121**, 401–423.

STELL, W. K. (1975) Horizontal cell axons and axon terminals in goldfish retina. *Journal of Comparative Neurology* **159**, 503–520.

STELL, W. K. & HAROSI, F. I. (1975) Cone structure and visual pigment content in the retina of the goldfish. *Vision Research* **16**, 647–657.

STELL, W. K. & LIGHTFOOT, D. O. (1975) Color-specific interconnections of cones and horizontal cells in the retina of the goldfish. *Journal of Comparative Neurology* **159**, 473–502.

STENKAMP, D. L., POWERS, M. K., CARNEY, L. H. & CAMERON, D. A. (2001) Evidence for two distinct mechanisms of neurogenesis and cellular pattern formation in regenerated goldfish retinas. *Journal of Comparative Neurology* **431**, 363–381.

STUDHOLME, K. M. & YAZULLA, S. (1988) Localization of GABA and glycine in goldfish retina by electron microscopic postembedding immunocytochemistry: Improved visualization of synaptic structures with LR white resin. *Journal of Neurocytology* **17**, 859–870.

TUMOSA, N., ECKENSTEIN, F. & STELL, W. K. (1984) Immunocytochemical localization of putative cholinergic neurons in the goldfish retina. *Neuroscience Letters* **48**, 255–259.

TUMOSA, N. & STELL, W. K. (1986) Choline acetyltransferase immunoreactivity suggests that ganglion cells in the goldfish retina are not cholinergic. *Journal of Comparative Neurology* **244**, 267–275.

VAN HAESENDONCK, E., MARC, R. E. & MISSOTTEN, L. (1993) New aspects of dopaminergic interplexiform cell organization in the goldfish retina. *Journal of Comparative Neurology* **333**, 503–518.

VERWEIJ, J., KAMERMANS, M., NEGISHI, K. & SPEKREIJSE, H. (1998) GABA sensitivity of spectrally classified horizontal cells in goldfish retina. *Visual Neuroscience* **15**, 77–86.

WAGNER, H. J. & WAGNER, E. (1988) Amacrine cells in the retina of a teleost fish, the roach (Rutilus rutilus): A Golgi study on differentiation and layering. *Philosophical Transactions of the Royal Society of London—Series B: Biological Sciences* **321**, 263–324.

WATT, C. B., KALLONIATIS, M., JONES, B. W. & MARC, R. E. (2000) Studies examining the neurotransmitter properties of horizontal cell populations in the goldfish retina. *Investigative Ophthalmology and Visual Science* **41**, S943.

YANG, C. Y. & YAZULLA, S. (1988) Localization of putative GABAergic neurons in the larval tiger salamander retina by immunocytochemical and autoradiographic methods. *Journal of Comparative Neurology* **277**, 96–108.

YAZULLA, S. (1981) Gamma aminobutyric-acid-ergic synapses in the goldfish carassius-auratus retina an auto radiographic study of tritium labeled muscimol and tritium labeled gamma aminobutyric-acid binding. *Journal of Comparative Neurology* **200**, 83–94.

YAZULLA, S. (1991) The mismatch problem for gabaergic amacrine cells in goldfish retina resolution and other issues. *Neurochemical Research.*

YAZULLA, S., MOSINGER, J. & ZUCKER, C. (1984) Two types of pyriform Ab amacrine cells in the goldfish retina: An EM analysis of [^{3}H]GABA uptake and somatostatin-like immunoreactivity. *Brain Research* **321**, 352–356.

YAZULLA, S. & STUDHOLME, K. M. (1991) Glycinergic interplexiform cells make synaptic contact with amacrine cell bodies in goldfish retina. *Journal of Comparative Neurology* **310**, 1–10.

YAZULLA, S., STUDHOLME, K. & WU, J. Y. (1986) Comparative distribution of 3H-GABA uptake and GAD immunoreactivity in goldfish retinal amacrine cells: A double-label analysis. *Journal of Comparative Neurology* **244**, 149–162.

YAZULLA, S., STUDHOLME, K. M. & WU, J. Y. (1987) GABAergic input to the synaptic terminals of mb1 bipolar cells in the goldfish retina. *Brain Research* **411**, 400–405.

YAZULLA, S., STUDHOLME, K. M. & ZUCKER, C. L. (1985) Synaptic organization of substance P-like immunoreactive amacrine cells in goldfish retina. *Journal of Comparative Neurology* **231**, 232–238.

YAZULLA, S. & ZUCKER, C. L. (1988) Synaptic organization of dopaminergic interplexiform cells in the goldfish retina. *Visual Neuroscience* **1**, 13–29.